SAFETY SYMBOLS	HAZARD	EXAMPLES	PRECAUTION	REMEDY
DISPOSAL	Special disposal procedures need to be followed.	certain chemicals, living organisms	Do not dispose of these materials in the sink or trash can.	Dispose of wastes as directed by your teacher.
BIOLOGICAL	Organisms or other biological materials that might be harmful to humans	bacteria, fungi, blood, unpreserved tissues, plant materials	Avoid skin contact with these materials. Wear mask or gloves.	Notify your teacher if you suspect contact with material. Wash hands thoroughly.
EXTREME TEMPERATURE	Objects that can burn skin by being too cold or too hot	boiling liquids, hot plates, dry ice, liquid nitrogen	Use proper protection when handling.	Go to your teacher for first aid.
SHARP OBJECT	Use of tools or glassware that can easily puncture or slice skin	razor blades, pins, scalpels, pointed tools, dissecting probes, broken glass	Practice common-sense behavior and follow guidelines for use of the tool.	Go to your teacher for first aid.
FUME	Possible danger to respiratory tract from fumes	ammonia, acetone, nail polish remover, heated sulfur, moth balls	Make sure there is good ventilation. Never smell fumes directly. Wear a mask.	Leave foul area and notify your teacher immediately.
ELECTRICAL	Possible danger from electrical shock or burn	improper grounding, liquid spills, short circuits, exposed wires	Double-check setup with teacher. Check condition of wires and apparatus.	Do not attempt to fix electrical problems. Notify your teacher immediately.
IRRITANT	Substances that can irritate the skin or mucus membranes of the respiratory tract	pollen, moth balls, steel wool, fiberglass, potassium permanganate	Wear dust mask and gloves. Practice extra care when handling these materials.	Go to your teacher for first aid.
CHEMICAL	Chemicals that can react with and destroy tissue and other materials	bleaches such as hydrogen peroxide; acids such as sulfuric acid, hydrochloric acid; bases such as ammonia, sodium hydroxide	Wear goggles, gloves, and an apron.	Immediately flush the affected area with water and notify your teacher.
TOXIC	Substance may be poisonous if touched, inhaled, or swallowed	mercury, many metal compounds, iodine, poinsettia plant parts	Follow your teacher's instructions.	Always wash hands thoroughly after use. Go to your teacher for first aid.
OPEN FLAME	Open flame may ignite flammable chemicals, loose clothing, or hair	alcohol, kerosene, potassium permanganate, hair, clothing	Tie back hair. Avoid wearing loose clothing. Avoid open flames when using flammable chemicals. Be aware of locations of fire safety equipment.	Notify your teacher immediately. Use fire safety equipment if applicable.

Eye Safety
Proper eye protection should be worn at all times by anyone performing or observing science activities.

Clothing Protection
This symbol appears when substances could stain or burn clothing.

Animal Safety
This symbol appears when safety of animals and students must be ensured.

Radioactivity
This symbol appears when radioactive materials are used.

New York, New York Columbus, Ohio Woodland Hills, California Peoria, Illinois

Glencoe Science

Chemistry

Student Edition
Teacher Wraparound Edition
Interactive Teacher Edition CD-ROM
Interactive Lesson Planner CD-ROM
Lesson Plans
Content Outline for Teaching
Dinah Zike's Teaching Science with Foldables
Directed Reading for Content Mastery
Foldables: Reading and Study Skills
Assessment
- *Chapter Review*
- *Chapter Tests*
- *ExamView Pro Test Bank Software*
- *Assessment Transparencies*
- *Performance Assessment in the Science Classroom*
- *The Princeton Review Standardized Test Practice Booklet*

Directed Reading for Content Mastery in Spanish
Spanish Resources
English/Spanish Guided Reading Audio Program
Reinforcement
Enrichment
Activity Worksheets
Section Focus Transparencies
Teaching Transparencies
Laboratory Activities
Science Inquiry Labs
Critical Thinking/Problem Solving
Reading and Writing Skill Activities
Mathematics Skill Activities
Cultural Diversity
Laboratory Management and Safety in the Science Classroom
Mindjogger Videoquizzes and Teacher Guide
Interactive Explorations and Quizzes CD-ROM with Presentation Builder
Vocabulary Puzzlemaker Software
Cooperative Learning in the Science Classroom
Environmental Issues in the Science Classroom
Home and Community Involvement
Using the Internet in the Science Classroom

"Study Tip," "Test-Taking Tip," and the "Test Practice" features in this book were written by The Princeton Review, the nation's leader in test preparation. Through its association with McGraw-Hill, The Princeton Review offers the best way to help students excel on standardized assessments.

The Princeton Review is not affiliated with Princeton University or Educational Testing Service.

Glencoe/McGraw-Hill
*A Division of The **McGraw-Hill** Companies*

The "Science and Society" and the "Science and History" features that appear in this book were designed and developed by TIME for Kids, a division of TIME Magazine.

Cover Images: Water droplets condense on the surface of a flask that has been cooled by a chemical reaction.

Send all inquires to:
Glencoe/McGraw-Hill
8787 Orion Place
Columbus, OH 43240

ISBN 0-07-825596-1
Printed in the United States of America.
1 2 3 4 5 6 7 8 9 10 027/111 06 05 04 03 02 01

Authors

Eric Werwa, PhD
Department of Physics and Astronomy
Otterbein College
Westerville, Ohio

Dinah Zike
Educational Consultant
Dinah-Might Activities, Inc.
San Antonio, Texas

Consultants

Content

Lisa McGaw
Science Teacher
Hereford High School
Hereford, Texas

Lee Meadows, PhD
UAL Birmingham Education Department
Birmingham, Alabama

Safety

Aileen Duc, PhD
Science II Teacher
Hendrick Middle School
Plano, Texas

Sandra West, PhD
Associate Professor of Biology
Southwest Texas State University
San Marcos, Texas

Reading

Rachel Swaters
Science Teacher
Rolla Middle School
Rolla, Missouri

Math

Michael Hopper, DEng
Manager of Aircraft Certification
Raytheon Company
Greenville, Texas

Reviewers

Sharla Adams
McKinney High School North
McKinney, Texas

Desiree Bishop
Baker High School
Mobile, Alabama

Nora M. Prestinari Burchett
Saint Luke School
McLean, Virginia

Maria Kelly
St. Leo School
Fairfax, Virginia

H. Keith Lucas
Stewart Middle School
Fort Defiance, Virginia

Linda Melcher
Woodmont Middle School
Piedmont, South Carolina

Annette Parrott
Lakeside High School
Atlanta, Georgia

Series Activity Testers

José Luis Alvarez, PhD
Math/Science Mentor Teacher
Yseleta ISD
El Paso, Texas

Nerma Coats Henderson
Teacher
Pickerington Jr. High School
Pickerington, Ohio

Mary Helen Mariscal-Cholka
Science Teacher
William D. Slider Middle School
El Paso, Texas

José Alberto Marquez
TEKS for Leaders Trainer
Yseleta ISD
El Paso, Texas

Science Kit and Boreal Laboratories
Tonawanda, New York

CONTENTS

Contents

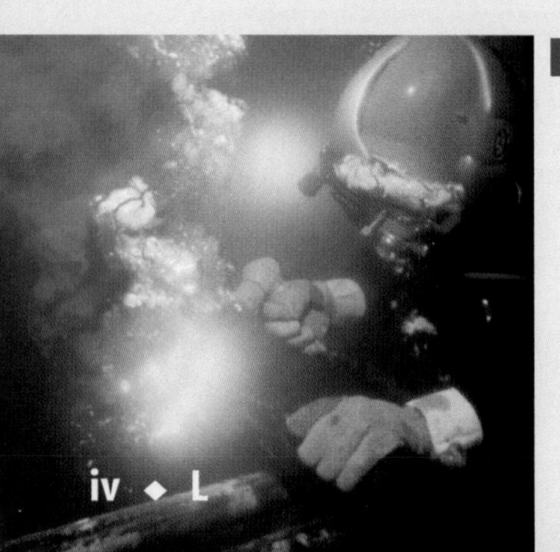

CHAPTER 4 Carbon Chemistry—94

Field Guide

Skill Handbooks—132

Reference Handbook

English Glossary—159

Spanish Glossary—163

Index—166

Interdisciplinary Connections/Activities

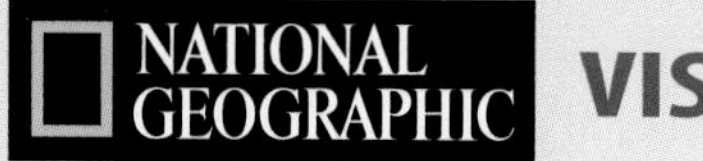

VISUALIZING

TIME SCIENCE AND Society

TIME SCIENCE AND HISTORY

Science and Language Arts

Science Stats

Full Period Labs

Mini LAB

EXPLORE ACTIVITY

Activities/Science Connections

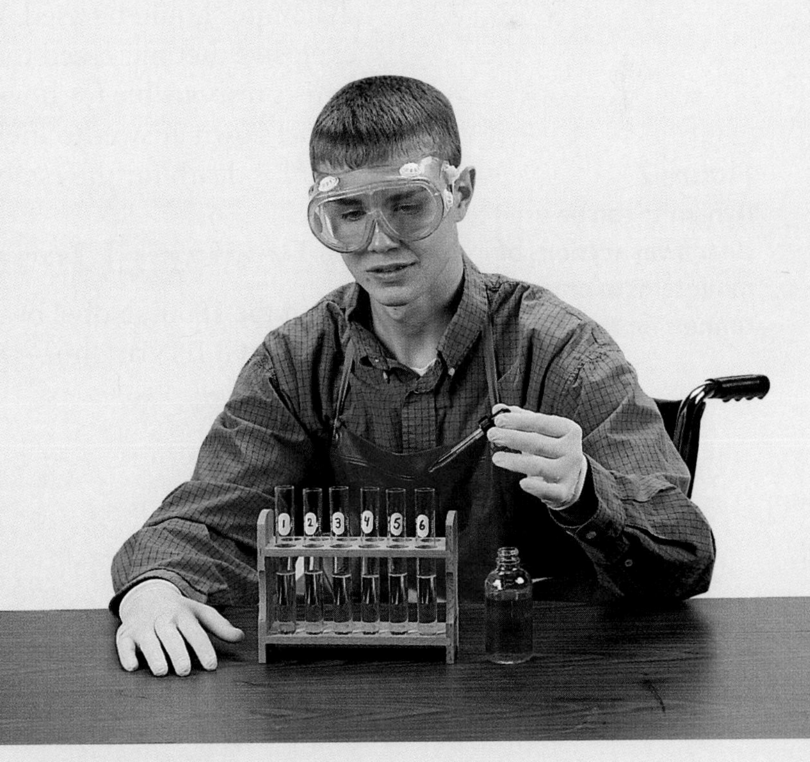

Feature Contents

Limits of Science

Alfred Nobel, Dynamite, and Peace

Alfred Nobel is best known for the invention of dynamite and the establishment of the Nobel prize—an award given to those whose efforts in the fields of physics, chemistry, medicine, literature, economics, or peace have benefited humanity. These two seemingly opposite acts—the invention of a deadly explosive and the establishment of a prize that promotes, among other things, peace—emphasize the power and the limitations of science.

Science enabled Nobel to create a superior explosive, but science could not answer other important questions, such as how dynamite should be used. Was it ethical for Nobel to invent an explosive that increased the killing power of weapons? Are scientists responsible for how their discoveries are used? Perhaps Nobel's own answer to these questions was to bestow money after his death for the establishment of the Nobel prize.

Figure 1
Alfred Nobel (1833–1896) invented both dynamite and the Nobel prize.

A Powerful Invention

In the 1850s, Nobel began experimenting with nitroglycerin (ni troh GLIHS or ohn)—a powerful liquid explosive. Since it exploded unpredictably, it was considered too dangerous for widespread use. Nobel decided to find a way to make nitroglycerin safer to handle. Nobel called his invention dynamite.

Figure 2
Dynamite can be used to clear away sections of mountains in order to build tunnels for roads and trains.

Nobel intended dynamite to be used as a construction tool. It was more powerful than gunpowder—the most common explosive used in construction at the time. Indeed, dynamite helped reduce the cost of blasting rocks, which is essential for building tunnels and canals. However, military leaders were also interested in Nobel's discovery. Only a few years after its invention, dynamite was used to kill soldiers in the Franco-Prussian War (1870–1871), a conflict between France and the German state of Prussia.

Nobel had not intended dynamite to be used as a weapon. Still, he became rich from selling dynamite to armies as well as to construction companies. He also later invented other explosives specifically for use in missiles, torpedoes, and cannons.

Some biographers claim that Nobel believed scientists are not responsible for how their discoveries are used. Others assert that Nobel founded a prize that promotes peace to counteract the harm done by his contribution to weapons. Clearly, science can't answer all the questions about scientists's accountability for their discoveries.

Figure 3
Dynamite was originally intended to be used in construction. This tunnel was created with dynamite.

Figure 4
During the Franco-Prussian War (1870–1871), dynamite was used as a weapon.

Science

Science is the process of gaining knowledge by asking questions and seeking the answers to these questions. To answer questions, scientists use scientific methods. They include identifying a question, forming and testing a hypothesis, analyzing results, and drawing conclusions. Alfred Nobel invented dynamite by beginning with the question "How can nitroglycerin be made more stable and therefore safer to handle?" After forming a hypothesis that nitroglycerin would be more stable if mixed with another substance, he tested several materials and finally found a safer mixture. For a question to be scientific, it must be testable. Scientific conclusions can change as more information is gained.

Figure 4
Before there were faces on Mount Rushmore, there was unshaped rock. Dynamite was used to form the noses of each face. More delicate methods were used to complete other features.

The Power of Science

Alfred Nobel's invention has benefited human-kind in countless ways. The Panama Canal, Mount Rushmore, and many tunnels and mines were built with the aid of dynamite. Dynamite also is used to build bridge supports and roadways. It can break up dangerous ice and logs jams and it can quickly and safely reduce large buildings to rubble. Police departments use dynamite to detonate suspicious packages. Fire departments use it to put out oil well fires. The explosion of the dynamite requires a huge amount of oxygen and suffocates the fire.

Figure 5
In demolitions, explosions are carefully placed to ensure that the building collapses inward.

The Limits of Science

Using scientific methods is the best way to learn about how the world works, but science has its limitations. Scientists are sometimes unable to answer a question or solve a problem because they lack the necessary tools. This is often a temporary limitation because once the required tools are developed, science often provides answers. For example, scientists were unaware of the existence of Jupiter's moons before the telescope was invented.

Figure 6
When the world was presented with Dolly, a sheep produced as a result of cloning, many ethical questions were raised about future applications of this new biotechnology.

What Science Can't Answer

Science can't give answers to questions that are not testable or that can't be measured or observed. Three major areas in which science can't provide answers are questions about morality, values, and spirituality.

The idea that it was immoral for Nobel to sell dynamite to armies, for example, is not a scientific idea because it can't be scientifically tested. Similarly, science can't answer opinion questions about values like the modern-day question of how far advances in the field of cloning should be taken. Should scientists use the new techniques developed to clone a human being? Any possible answer is a matter of opinion and therefore can't be measured or tested. Finally, science can't answer questions about spiritual matters because they are unable to be observed, measured, or tested.

Science and Responsibility

Although science can't answer questions about ethics and values, it can provide facts that may help people to make informed decisions. Being familiar with facts on all sides of an issue and careful consideration of what the possible positive and negative effects might be to an individual, a society, or the environment can help people decide upon a course of action.

Each new scientific discovery brings new questions. Some of these questions concern ethical matters that can't be answered by science alone. Find out about a recipient of the Nobel Prize for Chemistry. What did the person do to win this honor? What ethical questions arise from his or her work?

CHAPTER 1

Chemical Bonds

The stadium lights, the lights of the city, and the blimp hovering over the stadium all have something in common—they use gases that are members of the same element family. The blimp contains helium and the lights might contain argon, krypton, or xenon. In this chapter, you'll learn about the unique properties of this element family along with the properties of other element families. You'll also learn how electrons can be lost, gained, and shared by atoms to form the chemical bonds that shape your world.

What do you think?

Science Journal Look at this photo with a classmate. What do you think these are? Here's a hint: *They're smaller than grains of dust, but much more orderly.* Write down your best guess in your Science Journal.

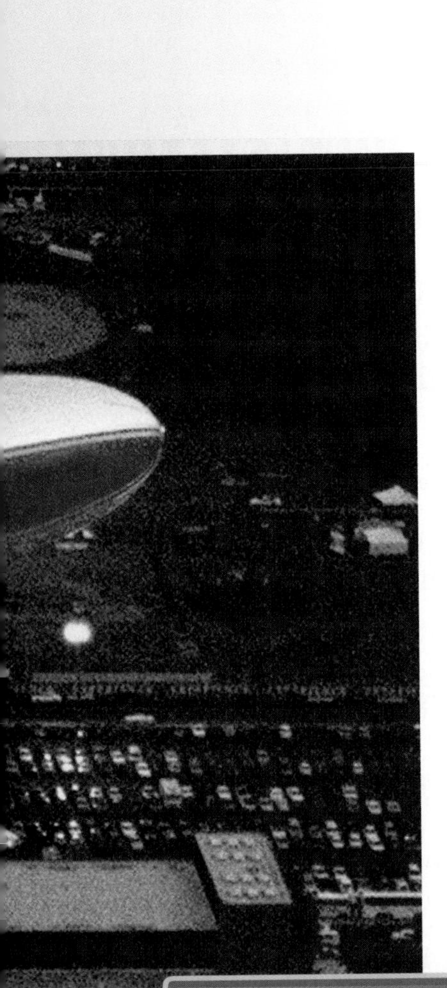

It's time to clean out your room—again. Where do all these things come from? Some are made of cloth and some of wood. The books are made of paper and an endless array of things are made of plastic. Fewer than 100 different kinds of naturally occurring elements are found on Earth. They combine to make all these different substances—but how? What makes elements form chemical bonds with other elements? The answer is in their electrons.

Model the energy of electrons

1. Pick up a paper clip with a magnet. Touch that paper clip to another paper clip and pick it up.
2. Continue picking up paper clips this way until you have a strand of them and no more will attach.
3. Then, gently pull off the paper clips one by one.

Observe

In your Science Journal, discuss which paper clip was easiest to remove and which was hardest. Was the clip that was easiest to remove closer to or farther from the magnet?

Before You Read

Making a Concept Map Study Fold **Make the following Foldable to help you organize information by diagramming ideas about chemical bonds.**

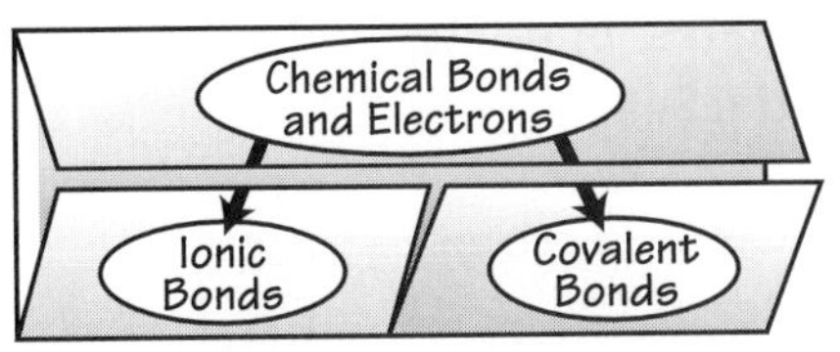

1. Place a sheet of paper in front of you so the long side is at the top. Fold the paper in half from top to bottom. Then, unfold the paper.
2. Fold the top and bottom in to the center fold. Fold from the left side to the right side. Unfold. Through the top thickness of paper, cut along the middle fold line of the bottom flap to form two tabs.
3. Draw an oval on the top flap and write *Chemical Bonds and Electrons* in the oval. Draw ovals on the bottom tabs and write *Ionic Bonds,* and *Covalent Bonds* in the ovals. Draw arrows from the top oval to each of the bottom ovals.
4. As you read the chapter, write information under each tab.

SECTION 1

Why do atoms combine?

As You Read

What You'll Learn

- **Identify** how electrons are arranged in an atom.
- **Determine** the energy of electrons in atoms.
- **Compare** how the arrangement of electrons in an atom is related to its place in the periodic table.

Vocabulary

electron cloud
electron dot diagram
chemical bond

Why It's Important

Chemical reactions take place all around you.

Atom Structure

When you look at your desk, you probably see it as something solid. You might be surprised to learn that all matter, even solids like wood and metal contain mostly empty space. How can this be? The answer is that although there might be little or no space between atoms, a lot of empty space lies within each atom.

At the center of every atom is a nucleus containing protons and neutrons. This nucleus represents most of the atom's mass. The rest of the atom is empty except for the atom's electrons, which are extremely small compared with the nucleus. Although the exact location of any one electron cannot be determined, the atom's electrons travel in an area of space around the nucleus called the **electron cloud.**

To visualize an atom, picture the nucleus as the size of a penny. In this case, electrons would be smaller than grains of dust and the electron cloud would extend outward as far as 20 football fields.

Electrons You might think that electrons resemble planets circling the Sun, but they are very different as you can see in **Figure 1.** First, planets have no charges, but the nucleus of an atom has a positive charge and electrons have negative charges.

Second, planets travel in predictable orbits—you can calculate exactly where one will be at any time. This is not true for electrons. Although electrons do travel in predictable areas, it is impossible to calculate the exact position of any one electron. Instead scientists use a model that predicts where an electron is most likely to be.

Figure 1
You can compare and contrast electrons with planets.

A Planets travel in well-defined paths, or orbits, around the Sun.

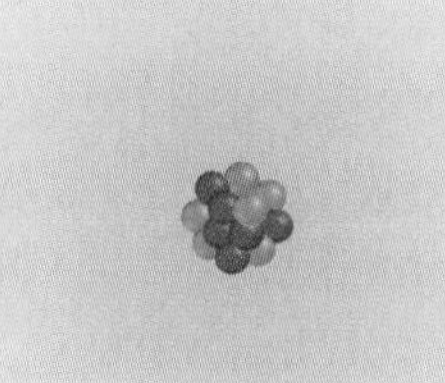

B Electrons travel around the nucleus. However, their paths are not well defined. They are more likely to be in the heavily shaded area than in the lightly shaded area, but they could be anywhere in the electron cloud.

Element Structure Each element has a different atomic structure consisting of a specific number of protons, neutrons, and electrons. The number of protons and electrons is always the same for a neutral atom of a given element. **Figure 2** shows a two-dimensional model of the electron structure of a lithium atom, which has three protons and four neutrons in its nucleus, and three electrons moving around its nucleus.

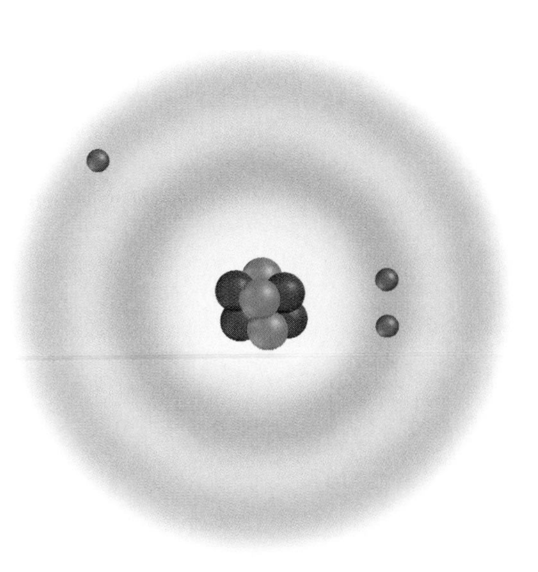

Electron Arrangement

The number and arrangement of electrons in the electron cloud of an atom are responsible for the physical and chemical properties of that element.

Electron Energy Although all the electrons in an atom are somewhere in the electron cloud, some electrons are closer to the nucleus than others. One model, which shows how the electrons might be arranged in energy levels around the nucleus, is shown in **Figure 3.** Energy levels, like layers in an onion, are a specific distance away from the nucleus. Each level represents a different amount of energy.

Figure 2
A neutral lithium atom has three positively charged protons, three negatively charged electrons, and four neutral neutrons.

Number of Electrons Each energy level can hold a specific number of electrons. The farther an energy level is from the nucleus, the more electrons it can hold. For example, the first energy level can hold one or two electrons, the second can hold up to eight, the third can hold up to 18, and the fourth energy level can hold a maximum of 32 electrons.

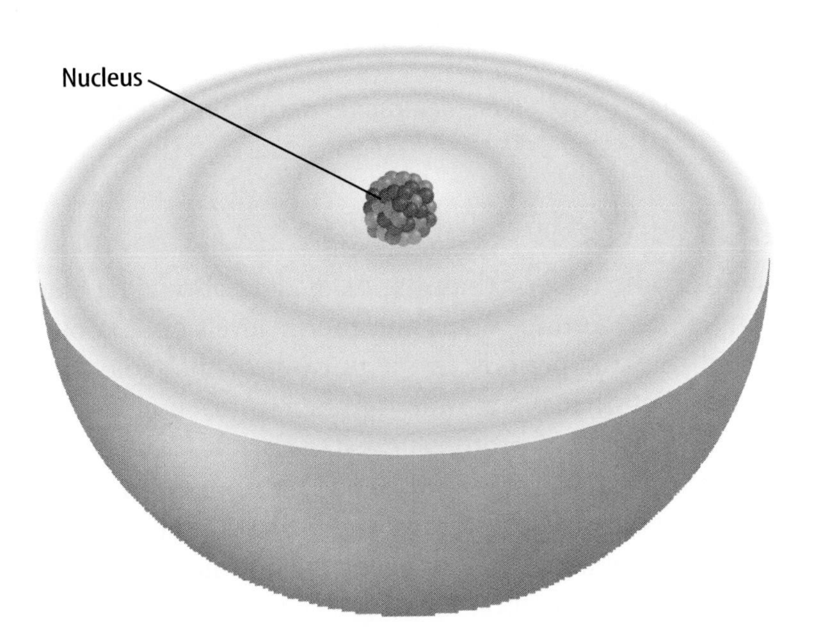

Figure 3
Electrons travel in three dimensions around the nucleus of an atom. The dark bands in this diagram show the energy levels where electrons are most likely to be found.

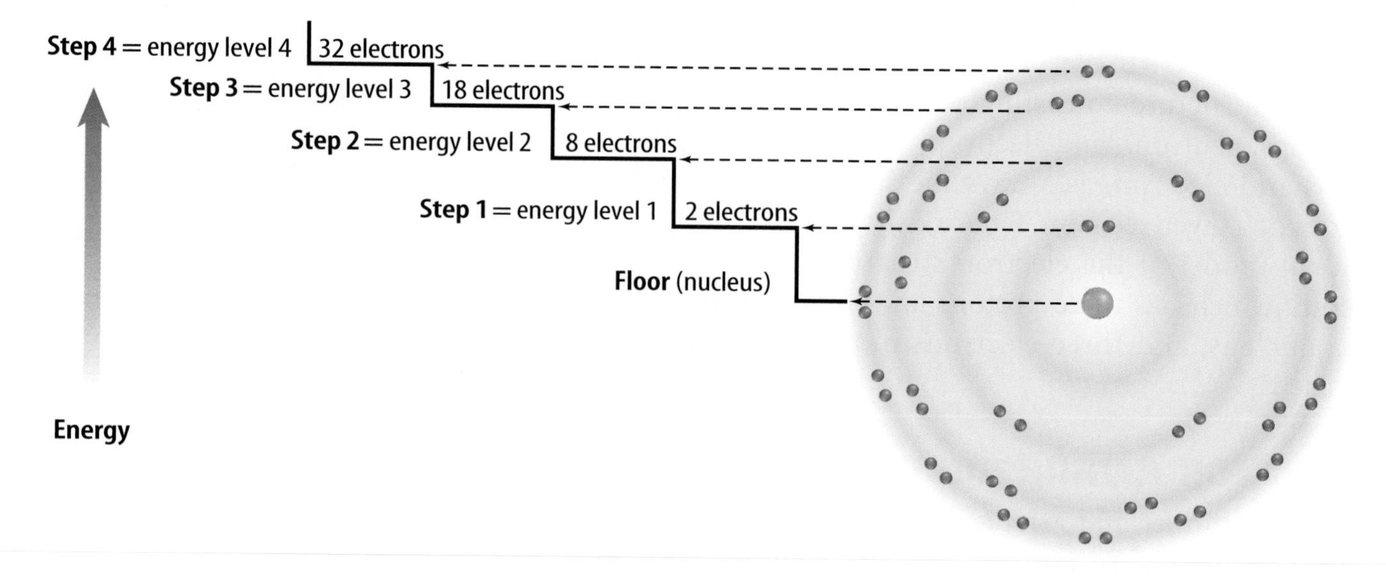

Figure 4
The farther an energy level is from the nucleus, the more electrons it can hold.

Energy Steps The stairway, shown in **Figure 4,** is a model that shows the number of electrons each energy level can hold in the electron cloud. Think of the nucleus as being at floor level. Electrons within an atom have different amounts of energy, represented by energy levels. These energy levels are represented by the stairsteps in Figure 4. Electrons in the level closest to the nucleus have the lowest amount of energy and are said to be in energy level one. Electrons farthest from the nucleus have the highest amount of energy and are the easiest to remove.

Recall the Explore Activity. It took more energy to remove the paper clip that was closest to the magnet than it took to remove the one that was farthest away. That's because the closer a paper clip was to the magnet, the stronger the magnet's attractive force was on the clip. Similarly, the closer a negatively charged electron is to the positively charged nucleus, the more strongly it is attracted to the nucleus. Therefore, removing electrons that are close to the nucleus takes more energy than removing those that are farther away from the nucleus.

Reading Check *What determines the amount of energy an electron has?*

Research Visit the Glencoe Science Web site at **science.glencoe.com** for more information about electrons. Communicate to your class what you learned.

Periodic Table and Energy Levels

The periodic table includes a lot of data about the elements and can be used to understand the energy levels also. Look at the horizontal rows, or periods, in the portion of the table shown in **Figure 5.** Recall that the atomic number for each element is the same as the number of protons in that element and that the number of protons equals the number of electrons in an electrically neutral atom. Therefore, you can determine the number of electrons in a neutral atom by looking at the atomic number written above each element symbol.

Electron Configurations

If you look at the periodic table shown in **Figure 5,** you can see that the elements are arranged in a specific order. The number of electrons in a neutral atom of the element increases by one from left to right across a period. For example, the first period consists of hydrogen with one electron and helium with two electrons in the energy level one. Recall from **Figure 4** that energy level one can hold up to two electrons. Therefore, helium's outer energy level is complete. Atoms with a complete outer energy level are stable. Therefore, helium is stable.

✔ Reading Check ***What term is given to the rows of the periodic table?***

The second period begins with lithium, which has three electrons—two in energy level one and one in energy level two. Lithium has one electron in its outer energy level. To the right of lithium is beryllium with two outer-level electrons, boron with three, and so on until you reach neon with eight.

Look again at **Figure 4.** You'll see that energy level two can hold up to eight electrons. Not only does neon have a complete outer energy level, but also this configuration of exactly eight electrons in an outer energy level is unusually stable. Therefore, neon is stable. The third period elements fill their outer energy levels in the same manner, ending with argon. Although energy level three can hold up to 18 electrons, argon has eight electrons in its outer energy level—a stable configuration. Each period in the periodic table ends with a stable element.

Figure 5
This portion of the periodic table shows the electron configurations of some elements. Count the electrons in each element and notice how the number increases across a period.

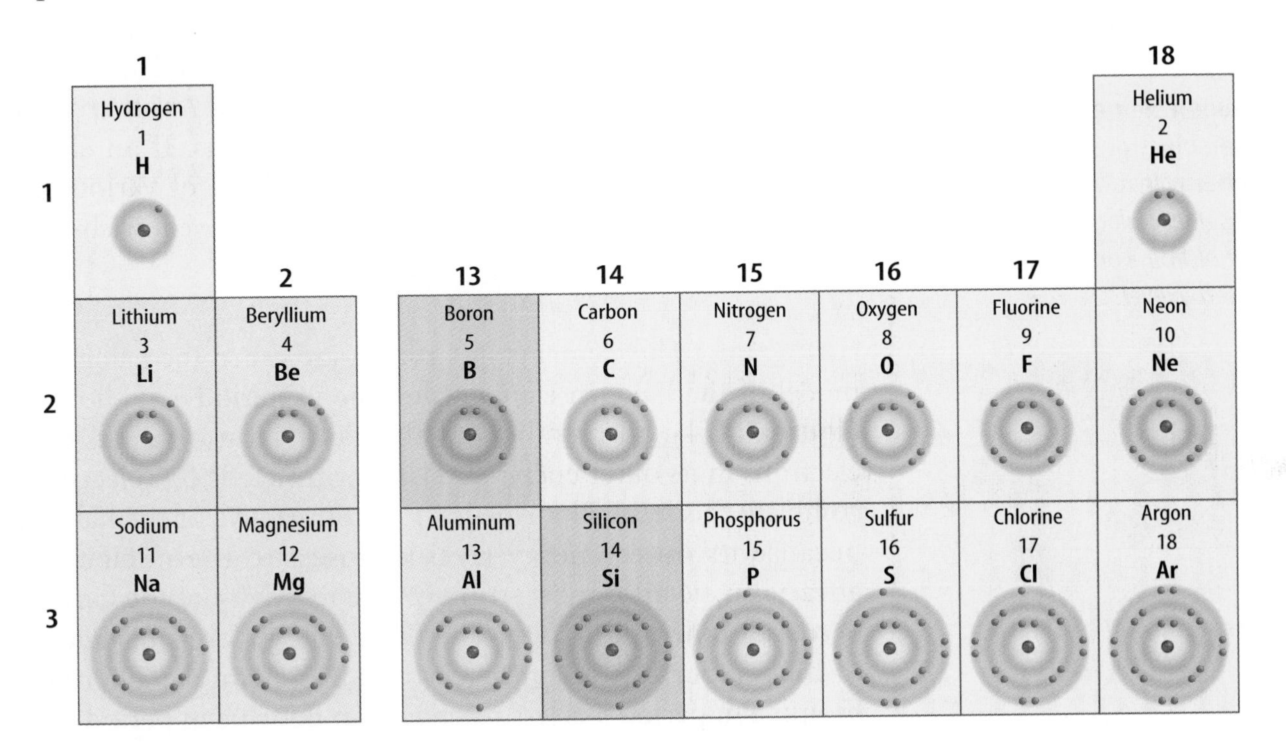

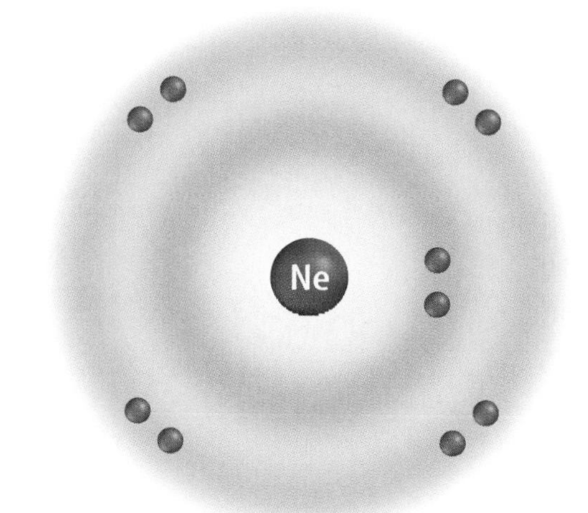

Figure 6
The noble gases are stable elements because their outer energy levels are complete or have a stable configuration of eight electrons like neon shown here.

Element Families

Elements can be divided into groups, or families. Each column of the periodic table in **Figure 5** contains one element family. Hydrogen is usually considered separately, so the first element family begins with lithium and sodium in the first column. The second family starts with beryllium and magnesium in the second column, and so on. Just as human family members often have similar looks and traits, members of element families have similar chemical properties because they have the same number of electrons in their outer energy levels.

It was this repeating pattern of properties that gave Russian chemist Dmitri Mendeleev the idea for his first periodic table in 1869. While listening to his family play music, he noticed how the melody repeated with increasing complexity. He saw a similar repeating pattern in the elements and immediately wrote down a version of the periodic table that looks much as it does today.

Noble Gases Look at the structure of neon in **Figure 6.** Neon and the elements below it in Group 8 have eight electrons in their outer energy levels. Their energy levels are stable, so they do not combine easily with other elements. Helium, with two electrons in its lone energy level, is also stable. At one time these elements were thought to be completely unreactive, and therefore became known as the inert gases. When chemists learned that these gases can be forced to react, their name was changed to noble gases. They are still the most stable element group.

This stability makes possible one widespread use of the noble gases—to protect filaments in lightbulbs. Another use of noble gases is to produce colored light in signs. If an electric current is passed through them they emit light of various colors—orange from neon, blue from argon, and yellowish-white from helium.

Figure 7
The halogen element fluorine has seven electrons in its outer energy level. *How many electrons does halogen family member bromine have in its outer energy level?*

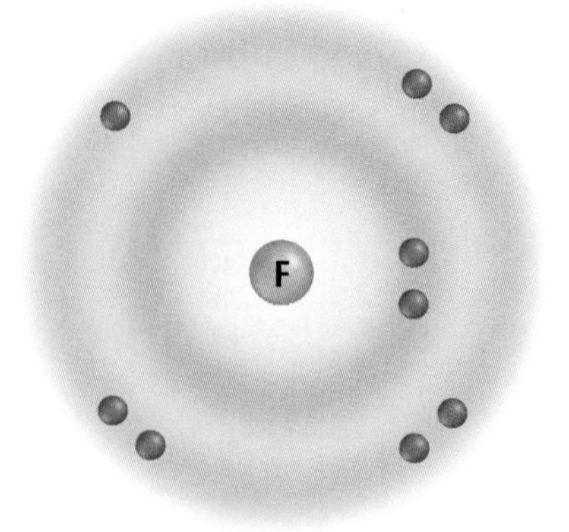

Halogens The elements in Group 7 are called the halogens. A model of the halogen element fluorine in period 2 is shown in **Figure 7.** Like all members of this family, fluorine has seven electrons in its outer energy level. Fluorine needs one electron to complete this level. Fluorine is the most reactive of the halogens because its outer energy level is closest to the nucleus and attracts an electron most strongly. The reactivity of the halogens decreases down the group. The outer energy levels of halogens in higher periods attract electrons less strongly. Therefore, bromine in period 4 is less reactive than fluorine in period 2.

Alkali Metals Next look at the element family in Group 1, called the alkali metals. The first members of this family, lithium and sodium, have one electron in their outer energy levels. You can see in **Figure 8** that potassium also has one electron in its outer level. Therefore, you can predict that the next family member, rubidium, does also. These electron arrangements are what determines how these metals react.

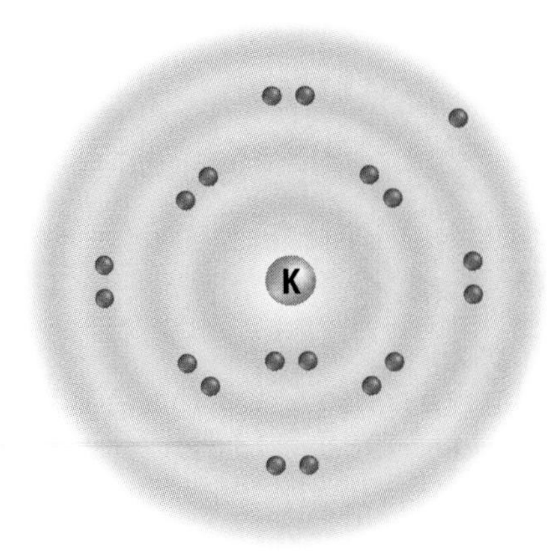

Figure 8
Potassium, like lithium and sodium, has only one electron in its outer level.

Reading Check *How many electrons do the alkali metals have in their outer energy levels?*

Like the halogens, the alkali metals form similar compounds. Alkali metals each have one outer energy level electron. It is this electron that is removed when alkali metals react. The easier it is to remove an electron, the more reactive the atom is. Unlike halogens, the reactivities of alkali metals increase down the group; that is, elements in the higher numbered periods are more reactive than elements in the lower numbered periods. This is because their outer energy levels are farther from the nucleus. Less energy is needed to remove an electron from an energy level that is farther from the nucleus than to remove one from an energy level that is closer to the nucleus. For this reason, cesium in period 6 loses an electron more readily and is more reactive than sodium in period 3.

Problem-Solving Activity

How does the periodic table help you identify properties of elements?

The periodic table displays information about the atomic structure of the elements. This information includes the properties, such as the energy level, of the elements. Can you identify an element if you are given information about its energy level? Use your ability to interpret the periodic table to find out.

Identifying the Problem

Recall that elements in a group in the periodic table contain the same number of electrons in their outer levels. The number of electrons increases by one from left to right across a period. Refer to **Figure 5.** Can you identify an unknown element or the group a known element belongs to?

Solving the Problem

1. An unknown element in Group 2 has a total number of 12 electrons and two electrons in its outer level. What is it?
2. Name the element that has eight electrons, six of which are in its outer level.
3. Silicon has a total of 14 electrons, four electrons in its outer shell, and three energy levels. What group does silicon belong to?
4. Three elements have the same number of electrons in their outer energy levels. One is oxygen. What might the other two be?

Drawing Electron Dot Diagrams

Procedure

1. Draw a periodic table that includes the first 18 elements—the elements from hydrogen through argon. Make each block a 3-cm square.
2. Fill in each block with the electron dot diagram of the element.

Analysis

1. What do you observe about the electron dot diagram of the elements in the same family?
2. Describe any changes you observe in the electron dot diagrams across a period.

Electron Dot Diagrams

The number of electrons in an element's outer energy level tells a lot about how that element behaves. Different atomic structures result in different physical properties such as color, hardness, and whether an element is a solid, liquid, or gas. Electronic structure also determines the chemical properties of an element; that is, how it behaves with other elements.

Drawing pictures of the energy levels and electrons in them takes time, especially when a large number of electrons are present. If you want to see at a glance how atoms of one element will behave, it is handy to have an easier way to represent the atoms and the electrons in their outer energy levels. You can do this with electron dot diagrams. An **electron dot diagram** is the symbol for the element surrounded by as many dots as there are electrons in its outer energy level. Only the outer energy level electrons are shown because these are what determines how an element can react.

How to Write Them How do you know how many dots to make? For Groups 1 and 2, and 13–18, you can use the periodic table or the portion of it shown in **Figure 5.** Group 1 has one outer electron. Group 2 has two. Group 13 has three, Group 14, four, and so on to Group 18. All members of Group 18 have stable outer energy levels. From neon down, they have eight electrons. Helium has only two electrons, but that is all that its single energy level can hold.

The dots are written in pairs on four sides of the element symbol. Start by writing one dot on the top of the element symbol, then work you way around, adding dots to the right, bottom, and left. Add a fifth dot to the top to make a pair. Continue in this manner until you reach eight dots to complete the level.

The process can be demonstrated by writing the electron dot diagram for the element nitrogen. First, write N—the element symbol for nitrogen. Then, find nitrogen in the periodic table and see what group it is in. It's in Group 15, so it has five electrons in its outer energy level. You can see the completed electron dot diagram for nitrogen in **Figure 9A.**

The electron dot diagram for iodine can be drawn the same way. The completed diagram is shown in **Figure 9B.**

Figure 9
These electron dot diagrams show only the electrons in the outer energy level.

A Nitrogen contains five electrons in its outer energy level.

B Iodine contains seven electrons in its outer energy level.

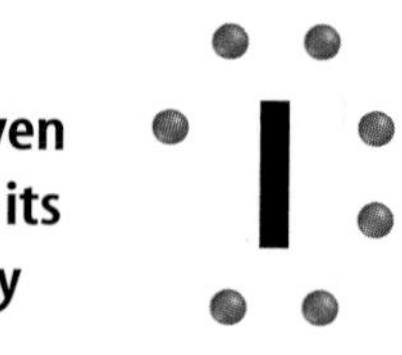

Figure 10
Some models are made by gluing pieces together. The glue that holds elements together in a chemical compound is the chemical bond.

Using Dot Diagrams Now that you know how to write electron dot diagrams for elements, you can use them to show how atoms bond with each other. A **chemical bond** is the force that holds two atoms together. Chemical bonds unite atoms in a compound much as glue unites the pieces of the model in **Figure 10.** Atoms bond with other atoms in such a way that each atom has a stable energy level. That is, their outer energy levels will resemble those of the noble gases.

Reading Check *What is a chemical bond?*

Section 1 Assessment

1. How many electrons does nitrogen have in its outer energy level? How many does bromine have?
2. How many electrons does oxygen have in its first energy level? Second energy level?
3. Which electrons in oxygen have the higher energy, those in the first energy level or those in the second?
4. Explain why elements in the same family have similar chemical properties.
5. **Think Critically** Atoms in a group of elements increase in size as you move down the columns in the periodic table. Explain why this is so.

Skill Builder Activities

6. **Classifying** Use the periodic table to organize the following elements into families: K, C, Sn, Li, F, Na, Pb, and I. Then, write the electron dot diagram for each element and compare them in each family. What can you conclude? **For more help, refer to the** Science Skill Handbook.
7. **Solving One-Step Equations** You can calculate the number of electrons in each energy level using the formula $2n^2$. Here, n is the number of the level and can have the values of 1, 2, 3, 4 and so on. Calculate the number of electrons in the first five energy levels. **For more help, refer to the** Math Skill Handbook.

SECTION

How Elements Bond

As You Read

What You'll Learn

- **Compare and contrast** ionic and covalent bonds.
- **Identify** the difference between polar and nonpolar covalent bonds.
- **Define** chemical shorthand.

Vocabulary

ion
ionic bond
compound
covalent bond
metallic bond
molecule
polar bond
chemical formula

Why It's Important

Chemical bonds join the atoms in the materials you use every day.

Ionic Bonds—Loss and Gain

When you put together the pieces of a jigsaw puzzle, they stay together only as long as you wish. When you pick up the completed puzzle, it falls apart. When elements are joined by chemical bonds, they do not readily fall apart. This is a good thing. What would happen if suddenly the salt you were shaking on your fries separated into sodium and chlorine? Atoms form bonds with other atoms using the electrons in their outer energy levels. They have four ways to do this—by losing electrons, by gaining electrons, by pooling electrons, or by sharing electrons with another element in a way that is advantageous to both.

Sodium is a soft, silvery metal as shown in **Figure 11A.** It can react violently when added to water or to chlorine. What makes sodium so reactive? If you look at a diagram of its energy levels in **Figure 11B,** you will see that sodium has only one electron in its outer level. Removing this electron empties this level and leaves the completed level below. By removing one electron, sodium's electron configuration becomes the same as that of the stable noble gas neon.

Chlorine forms bonds in a way that is the opposite of sodium—it completes its outer energy level by gaining an electron. When chlorine accepts an electron, its electron configuration becomes the same as that of the noble gas argon.

Figure 11
Sodium and chlorine react forming white crystalline sodium chloride.

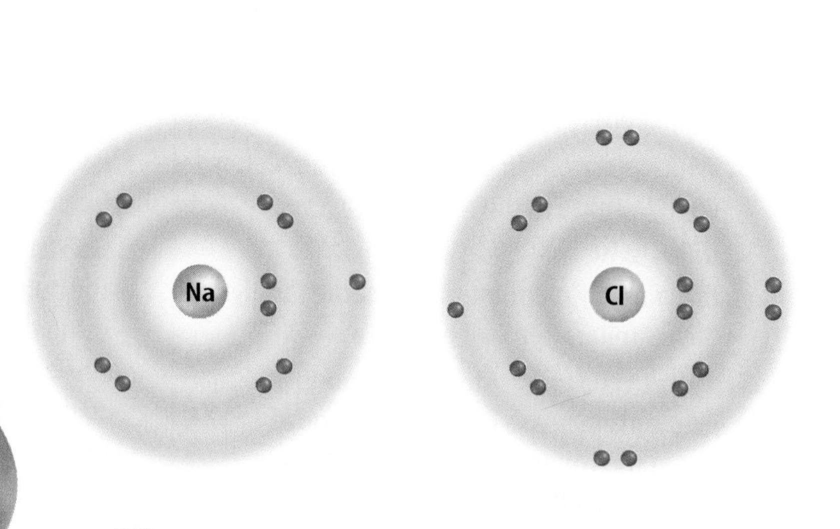

A Sodium is a silvery metal that can be cut with a knife. Chlorine is a greenish, poisonous gas.

B Their electronic structures show why they react.

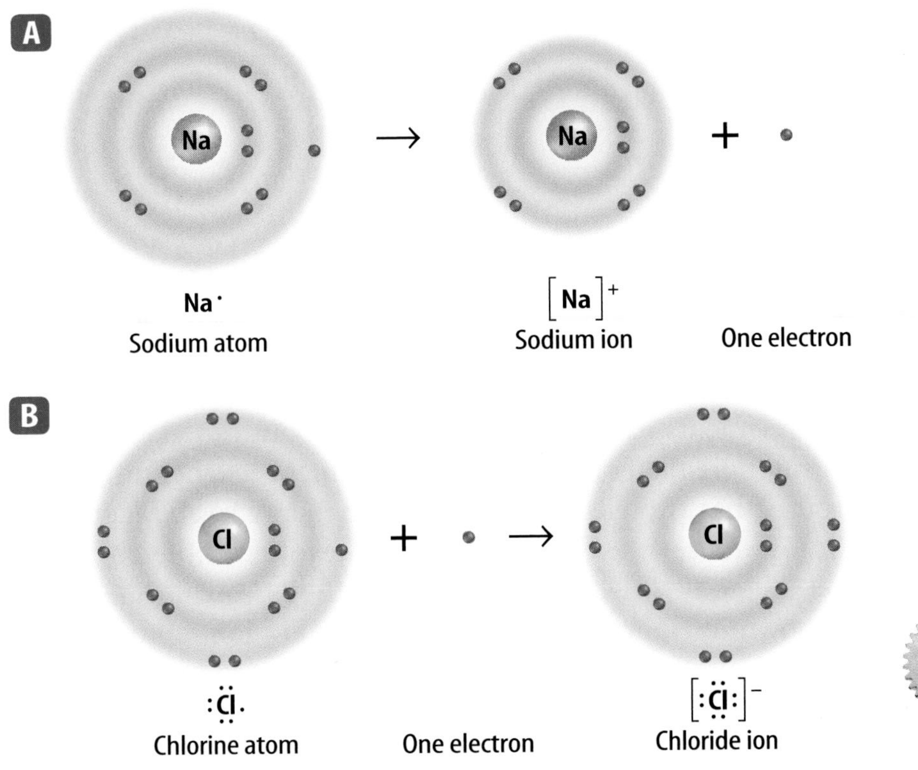

Figure 12
Ions form when elements lose or gain electrons. A Sodium loses one electron to become a Na^+ ion. B Chlorine gains one electron to become a Cl^- ion. The symbols in brackets represent ions.

Ions—A Question of Balance As you just learned, a sodium atom loses an electron and becomes more stable. But something else happens also. By losing an electron, the balance of electric charges changes. Sodium becomes positively charged because there is now one fewer electron than there are protons in the nucleus. In contrast, chlorine becomes an ion by gaining an electron. It becomes negatively charged because there is one more electron than there are protons in the nucleus.

An atom that is no longer neutral because it has lost or gained an electron is called an **ion** (I ahn). A sodium ion is represented by the symbol Na^+ and a chloride ion is represented by the symbol Cl^-. **Figure 12** shows dot diagrams for the two ions.

Bond Formation The positive sodium ion and the negative chloride ion are strongly attracted to each other. This attraction, which holds the ions close together, is a type of chemical bond called an **ionic bond.** In **Figure 13,** sodium and chloride ions form an ionic bond. The compound sodium chloride, or table salt, is formed. A **compound** is a pure substance containing two or more elements that are chemically bonded.

When ions dissolve in water, they separate. Because of their positive and negative charges, the ions can conduct an electric current. If wires are placed in such a solution and the ends of the wires are connected to a battery, the positive ions move toward the negative terminal and the negative ions move toward the positive terminal. This flow of ions completes the circuit.

$$Na^{\cdot} + \cdot\ddot{\underset{\cdot\cdot}{Cl}}: \rightarrow [Na]^+ [:\ddot{\underset{\cdot\cdot}{Cl}}:]^-$$

Figure 13
An ionic bond forms between atoms of opposite charges.

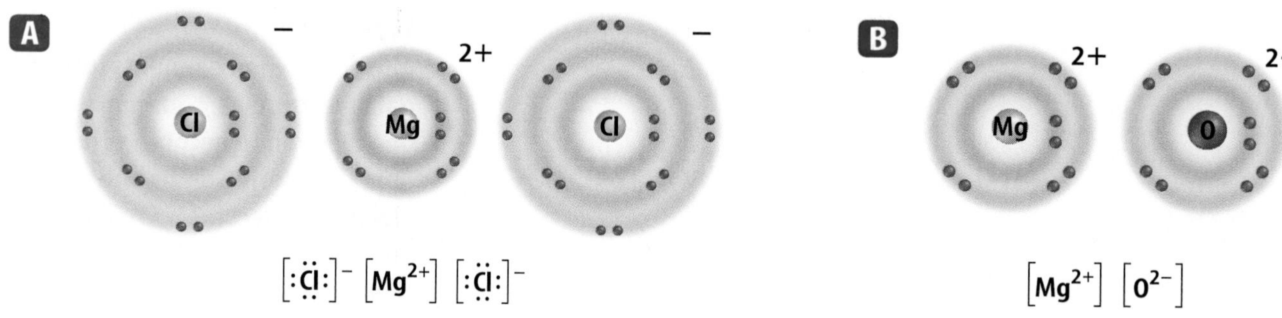

Magnesium chloride

Magnesium oxide

Figure 14
Magnesium has two electrons in its outer energy level. A If one electron is lost to each of two chlorine atoms, magnesium chloride forms. B If both electrons are lost to one oxygen atom, magnesium oxide forms.

More Gains and Losses You have seen what happens when elements gain or lose one electron, but can elements lose or gain more than one electron? The element magnesium, Mg, in Group 2 has two electrons in its outer energy level. Magnesium can lose these two electrons and achieve a completed energy level. These two electrons can be gained by two chlorine atoms. As shown in **Figure 14A,** a single magnesium ion represented by the symbol Mg^{2+} and two chloride ions are produced. The two negatively charged chloride ions are attracted to the positively charged magnesium ion forming ionic bonds. As a result of these bonds, the compound magnesium chloride is produced.

Some atoms, such as oxygen, need to gain two electrons to achieve stability. The two electrons released by one magnesium atom could be gained by a single atom of oxygen. When this happens, magnesium oxide (MgO) is formed, as shown in **Figure 14B.** Oxygen can form similar compounds with any positive ion from Group 2.

Metallic Bonding—Pooling

You have just seen how metal atoms form ionic bonds with atoms of nonmetals. Metals can form bonds with other metal atoms, but in a different way. In a metal, the electrons in the outer energy levels of the atoms are not held tightly to individual atoms. Instead, they move freely among all the ions in the metal, forming a shared pool of electrons, as shown in **Figure 15. Metallic bonds** form when metal atoms share their pooled electrons. This bonding affects the properties of metals. For example, when a metal is hammered into sheets or drawn into a wire, it does not break. Instead, layers of atoms slide over one another. The pooled electrons tend to hold the atoms together. Metallic bonding also is the reason that metals conduct electricity well. The outer electrons in metal atoms are held so weakly that they readily move from one atom to the next to transmit current.

Figure 15
In metallic bonding, the outer electrons of the silver ions are not attached to any one silver ion. This allows them to move and conduct electricity.

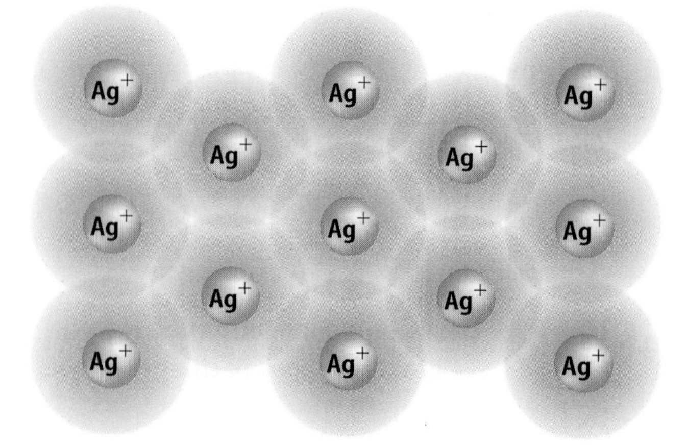

Covalent Bonds—Sharing

Some atoms are unlikely to lose or gain electrons because the number of electrons in their outer levels makes this difficult. For example, carbon has six protons and six electrons. Four of the six electrons are in its outer energy level. To obtain a more stable structure, carbon would either have to gain or lose four electrons. This is difficult because gaining and losing so many electrons takes so much energy. The alternative is sharing electrons.

The Covalent Bond Atoms of many elements become more stable by sharing electrons. The chemical bond that forms between atoms when they share electrons is called a **covalent** (koh VAY luhnt) **bond.** Shared electrons are attracted to the nuclei of both atoms. They move back and forth between the outer energy levels of each atom in the covalent bond. In this way each atom has a stable outer energy level some of the time. Covalently bonded compounds are called molecular compounds.

Reading Check *How do atoms form covalent bonds?*

The atoms in a covalent bond form a neutral particle, which contains the same numbers of positive and negative charges. The neutral particle formed when atoms share electrons is called a **molecule** (MAH lih kyewl). A molecule is the smallest unit of a molecular compound. You can see how molecules form by sharing electrons in **Figure 16.** Notice that no ions are involved because no electrons are gained or lost. Crystalline solids, such as sodium chloride, are not referred to as molecules, because their smallest units are ions.

Constructing a Model of Methane

Procedure

1. Using **circles of colored paper** to represent protons, neutrons, and electrons, build paper models of one carbon atom and four hydrogen atoms.
2. Use your models of atoms to construct a molecule of methane by forming covalent bonds. The methane molecule has four hydrogen atoms chemically bonded to one carbon atom.

Analysis

1. In the methane molecule, do the carbon and hydrogen atoms have the same arrangement of electrons as two noble gas elements? Explain your answer.
2. Does the methane molecule have a charge?

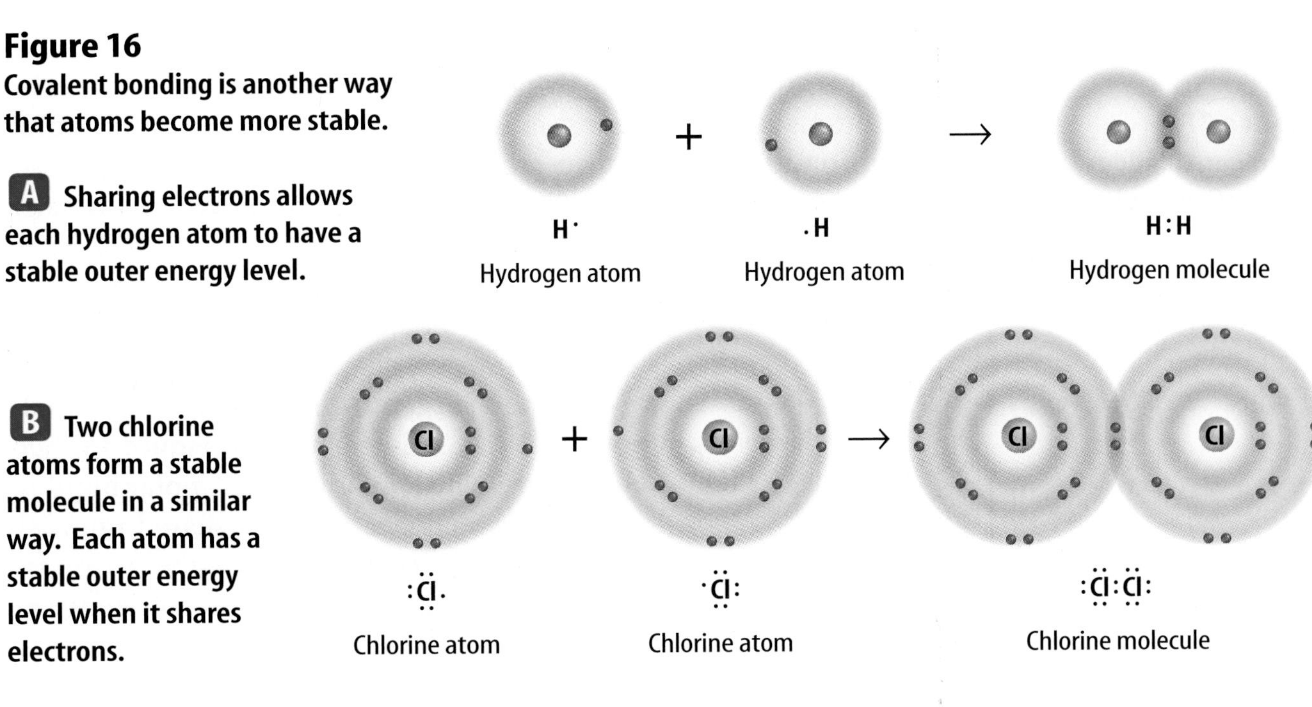

Figure 16
Covalent bonding is another way that atoms become more stable.

A **Sharing electrons allows each hydrogen atom to have a stable outer energy level.**

B **Two chlorine atoms form a stable molecule in a similar way. Each atom has a stable outer energy level when it shares electrons.**

Figure 17
Atoms can form covalent bonds by sharing two or even three electrons.

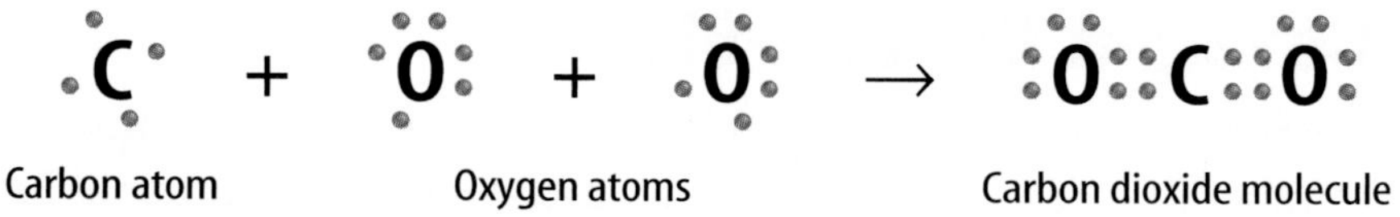

Carbon atom + Oxygen atoms → Carbon dioxide molecule

A **In carbon dioxide, carbon shares two electrons with each of two oxygen atoms forming two double bonds.**

Nitrogen atoms → Nitrogen molecule

B **Nitrogen shares three electrons in forming a triple bond with another nitrogen atom.**

Double and Triple Bonds Sometimes an atom shares more than one electron with another atom. In the molecule carbon dioxide, shown in **Figure 17A,** each of the oxygen atoms shares two electrons with the carbon atom. When two pairs of electrons are involved in a covalent bond, the bond is called a double bond. **Figure 17B** also shows the sharing of three pairs of electrons between two nitrogen atoms in the nitrogen molecule. When three pairs of electrons are shared by two atoms, the bond is called a triple bond. Nitrogen molecules make up about 80 percent of the air in the atmosphere.

Reading Check *How many pairs of electrons are shared in a double bond?*

Polar and Nonpolar Molecules

You have seen how atoms can share electrons and that they become more stable by doing so, but do they always share electrons equally? The answer is no. Some atoms have a greater attraction for electrons than others do. Chlorine, for example, attracts electrons more strongly than hydrogen does. When a covalent bond forms between hydrogen and chlorine, the shared pair of electrons tends to spend more time near the chlorine atom than the hydrogen atom.

This unequal sharing makes one side of the bond more negative than the other, like poles on a battery. This is shown in **Figure 18.** Such bonds are called polar bonds. A **polar bond** is a bond in which electrons are shared unevenly. The bonds between the oxygen atom and hydrogen atoms in the water molecule are another example of polar bonds.

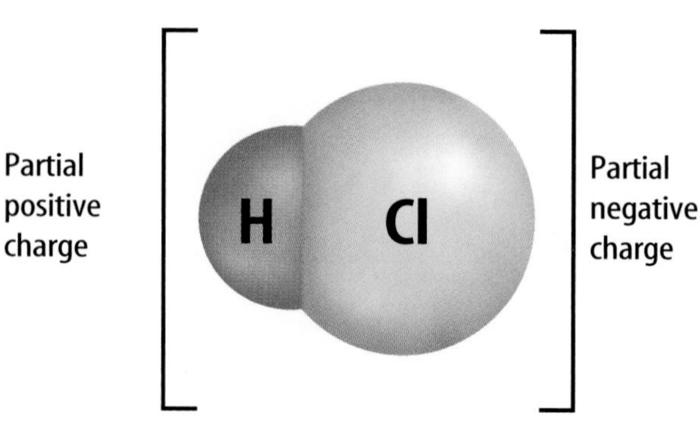

Figure 18
Hydrogen chloride is a polar covalent molecule.

The Polar Water Molecule Water molecules form when hydrogen and oxygen share electrons. **Figure 19A** shows how this sharing is unequal. The oxygen atom has a greater share of the electron pair—the oxygen end of a water molecule has a slight negative charge and the hydrogen end has a slight positive charge. Because of this, water is said to be polar—having two opposite ends or poles like a magnet.

When they are exposed to a negative charge, the water molecules line up like magnets with their positive ends facing the negative charge. You can see how they are drawn to the negative charge on the balloon in **Figure 19B.** Water molecules also are attracted to each other. This attraction between water molecules accounts for many of the physical properties of water.

Molecules that do not have these uneven charges are called nonpolar molecules. Because each element differs slightly in its ability to attract electrons, the only completely nonpolar bonds are bonds between atoms of the same element. One example of a nonpolar bond is the triple bond in the nitrogen molecule.

Like ionic compounds, some molecular compounds can form crystals, in which the basic unit is a molecule. Often you can see the pattern of the units in the shape of the crystal, as shown in **Figure 20.**

Research Visit the Glencoe Science Web site at **science.glencoe.com** for more information about polar molecules. Communicate to your class what you learn.

Figure 19
The water molecule is polar.

A **Two hydrogen atoms share electrons with one oxygen atom, but the sharing is unequal. The electrons are more likely to be closer to the oxygen than the hydrogens. The space-saving model shows how the charges are separated or polarized.**

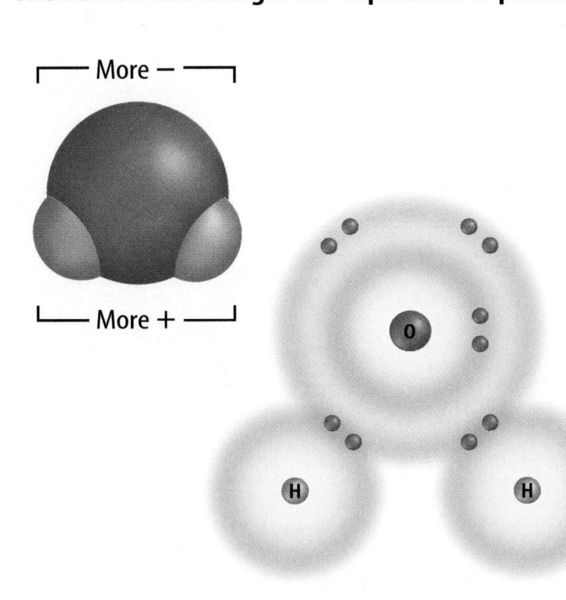

B **The positive ends of the water molecules are attracted to the negatively charged balloon, causing the stream of water to bend.**

NATIONAL GEOGRAPHIC **VISUALIZING CRYSTAL STRUCTURE**

Figure 20

Many solids exist as crystals. Whether tiny grains of table salt or big, chunky blocks of quartz you might find rock hunting, a crystal's shape is often a reflection of the arrangement of its particles. Knowing a solid's crystal structure helps researchers understand its physical properties. Some crystals with cubic and hexagonal shapes are shown here.

O

Si

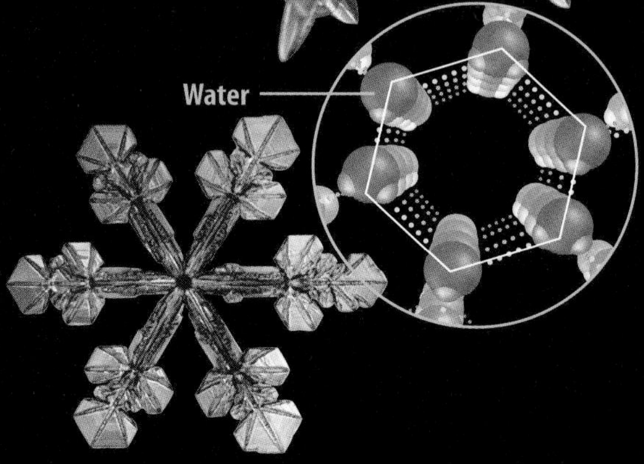

Water

HEXAGONAL Quartz crystals, above, are six sided, just as a snowflake, above right, has six points. This is because the molecules that make up both quartz and snowflakes arrange themselves into hexagonal patterns.

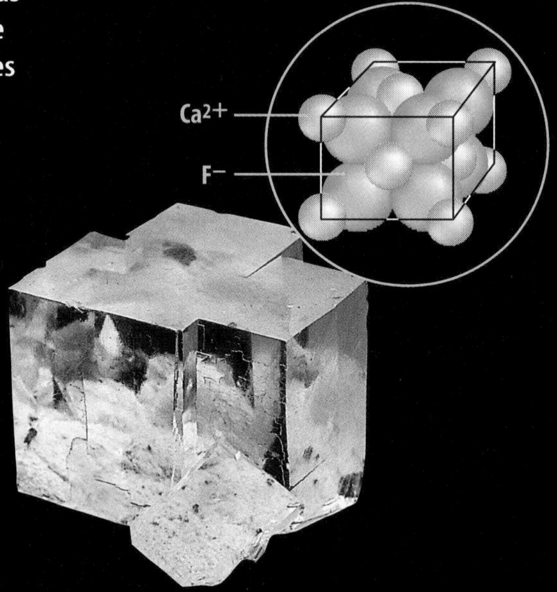

Ca^{2+}

F^-

Na^+

Cl^-

CUBIC Salt, left, and fluorite, above, form cube-shaped crystals. This shape is a reflection of the cube-shaped arrangement of the ions in the crystal.

Chemical Shorthand

In medieval times alchemists (AL kuh mists) were the first to explore the world of chemistry. Although many of them believed in magic and mystical transformations, alchemists did learn much about the properties of some elements. They even used symbols to represent them in chemical processes, some of which are shown in **Figure 21.**

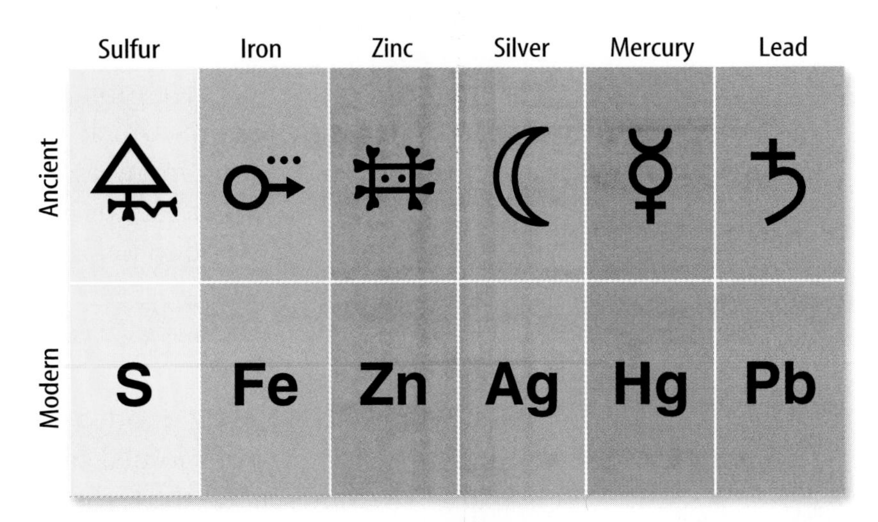

Figure 21
Alchemists used elaborate symbols to describe elements and processes. Modern chemical symbols are letters that can be understood all over the world.

Symbols for Atoms Modern chemists use symbols to represent elements, too. These symbols can be understood by chemists everywhere. Each element is represented by a one letter-, two letter-, or three-letter symbol. Many symbols are the first letters of the element's name, such as H for hydrogen and C for carbon. Others are the first letters of the element's name in another language, such as K for potassium, which stands for kalium, the Latin word for potassium.

Symbols for Compounds Compounds can be described using element symbols and numbers. For example, **Figure 22A** shows how two hydrogen atoms join together in a covalent bond. The resulting hydrogen molecule is represented by the symbol H_2. The small 2 after the H in the formula is called a subscript. *Sub* means "below" and *script* means "write," so a subscript is a number that is written below. The subscript 2 means that two atoms of hydrogen are in the molecule.

Figure 22
Chemical formulas show you the kind and number of atoms in a molecule.

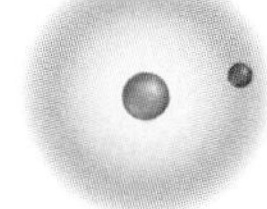

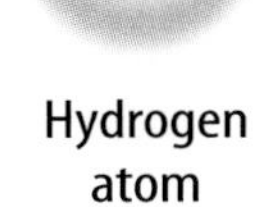
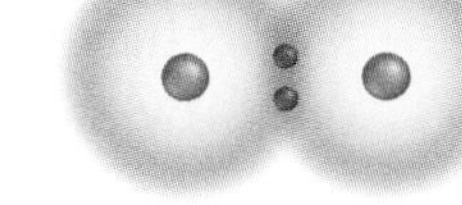

Hydrogen atom + Hydrogen atom → H_2 molecule

A The subscript 2 after the H indicates that the hydrogen molecule contains two atoms of hydrogen.

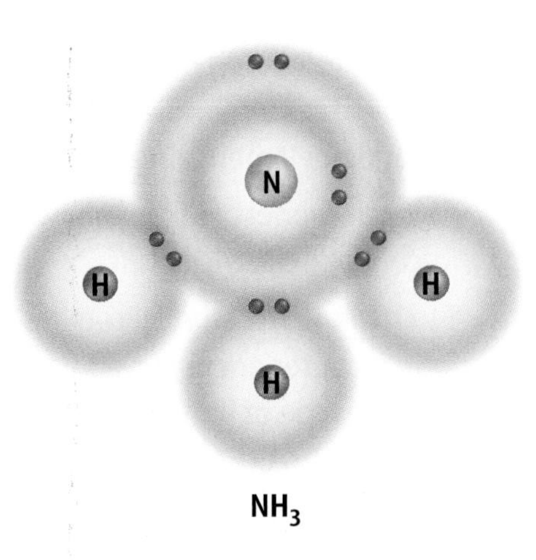

B The formula for ammonia, NH^3, tells you that the ratio is one nitrogen atom to three hydrogen atoms.

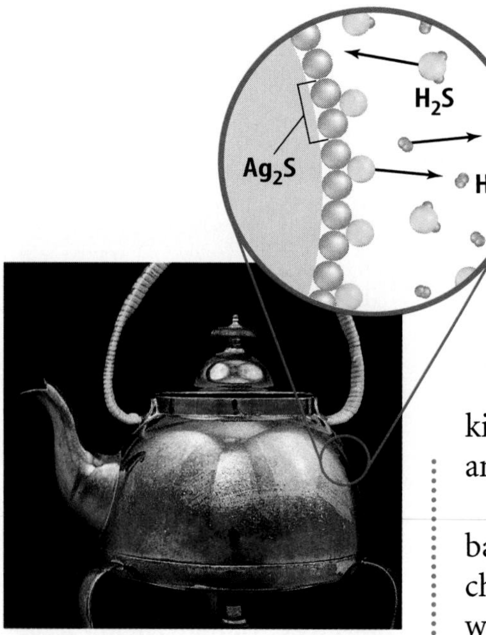

Figure 23
Silver tarnish is the compound silver sulfide, Ag_2S. The formula shows that two silver atoms are combined with one sulfur atom.

Chemical Formulas A **chemical formula** is a combination of chemical symbols and numbers that shows which elements are present in a molecule and how many atoms of each element are present. When no subscript is shown, the number of atoms is understood to be one.

Reading Check *What is a chemical formula and what does it tell you about the molecule?*

Covalently bonded molecules also can have more than one kind of atom joined together in covalent bonds. For example, ammonia has the formula NH_3, as shown in **Figure 22B.**

Now that you understand chemical formulas, you can look back at the other chemical compounds shown earlier in this chapter, and write their chemical formulas. For example, the water molecule shown in **Figure 19A** contains one oxygen atom and two hydrogen atoms, so its formula is H_2O.

The black tarnish that forms on silver, shown in **Figure 23,** is a compound made up of the elements silver and sulfur in the proportion of two atoms of silver to one atom of sulfur. If alchemists knew the composition of silver tarnish, how might they have written a formula for the compound? The modern formula for silver tarnish is Ag_2S. The formula tells you that it is a compound that contains two silver atoms and one sulfur atom.

Section 2 Assessment

1. Use the periodic table to decide whether lithium forms a positive or a negative ion. Does fluorine form a positive or a negative ion? Write the formula for the compound formed from these two elements.
2. What is the difference between a polar and a nonpolar bond?
3. How does a chemical formula indicate the ratio of elements in a compound?
4. What property of ions allows them to conduct electricity?
5. **Think Critically** Silicon has four electrons in its outer energy level. Based on this fact, what type of bond is silicon most likely to form with other elements? Explain.

Skill Builder Activities

6. **Predicting** Scientists use what they have learned to predict what they think will happen. Predict the type of bond that will form between the following pairs of atoms: carbon and oxygen, potassium and bromine, fluorine and fluorine. **For more help, refer to the Science Skill Handbook.**
7. **Using an Electronic Spreadsheet** Design a table using a spreadsheet to compare and contrast ionic, polar covalent, and nonpolar covalent bonds. Include a description, properties and examples of each. **For more help, refer to the Technology Skill Handbook.**

Activity

Ionic Compounds

Metals in Groups 1 and 2 often lose electrons and form positive ions. Nonmetals in Groups 15, 16, and 17 often gain electrons and become negative ions. How can compounds form between these five groups of elements?

What You'll Investigate

How do different atoms combine with each other to form compounds?

Materials

paper (8 different colors)
tacks (2 different colors)
corrugated cardboard
scissors

Goals

- **Construct** models of electron gain and loss.
- **Determine** formulas for the ions and compounds that form when electrons are gained or lost.

Safety Precautions

Procedure

1. Cut colored-paper disks 7-cm in diameter to represent the elements Li, S, Mg, O, Ca, Cl, Al, and I. Label each disk with one symbol.
2. Lay circles representing the atoms Li and S side by side on cardboard.
3. Choose colored thumbtacks to represent the outer electrons of each atom. Place the tacks evenly around the disks to represent the outer electron levels of the elements.
4. Move electrons from the metal atom to the nonmetal atom so that both elements achieve noble gas arrangements. If needed, cut additional paper disks to add more atoms of one element.
5. Write the formula for each ion formed when you shift electrons.

6. Repeat steps 2 through 6 to combine Mg and O, Ca and Cl, and Al and I.

Conclude and Apply

1. **Draw** electron dot diagrams for all of the ions produced.
2. **Identify** the noble gas elements having the same electron arrangements as the ions you made in this activity.
3. Why did you have to use more than one atom in some cases? Why couldn't you take more electrons from one metal atom or add extra ones to a nonmetal atom?

Communicating Your Data

Compare your compounds and dot diagrams with those of other students in your class. **For more help, refer to the Science Skill Handbook.**

Activity Design Your Own Experiment

Atomic Structure

As more information has become known about the structure of the atom, scientists have developed new models. Making your own model and studying the models of others will help you learn how protons, neutrons, and electrons are arranged in an atom.

Recognize the Problem

Can an element be identified based on a model that shows the arrangement of the protons, neutrons, and electrons of an atom?

Form a Hypothesis

Write a hypothesis that explains how your group will construct a model of an element that others will be able to identify.

Possible Materials

magnetic board
rubber magnetic strips
half-inch squares of paper
candy-coated peanuts
scissors
paper
marker
grapes
coins

Goals

- **Design** an experiment to create a model of a chosen element.
- **Observe** the models made by others in the class and identify the elements they represent.

Safety Precautions

WARNING: *Never eat any food used in a laboratory experiment. Wash hands thoroughly.*

Test Your Hypothesis

Plan

1. Choose an element from periods 2 or 3 of the periodic table. How can you determine the number of protons, neutrons, and electrons in an atom?
2. How can you show the difference between protons and neutrons? What materials will you use to represent the electrons of the atom? How will you represent the nucleus?
3. How will you model the arrangement of electrons in the atom? Will the atom have a charge? Is it possible to identify an atom by the number of protons it has?

Do

1. Make sure your teacher approves your plan before you proceed.
2. **Construct** your model. Then record your observations in your Science Journal and include a sketch.
3. **Construct** another model of a different element.
4. **Observe** the models made by your classmates. Identify the elements they represent.

Analyze Your Data

1. What elements did you identify using your hypothesis?
2. In a neutral atom, identify which particles always are present in equal numbers.
3. **Predict** what would happen to the charge on an atom if one of the electrons were removed.
4. What happens to an atom if two electrons are added? What happens to an atom if one proton and one electron are removed?
5. **Compare and contrast** your model with the electron cloud model of the atom. How is your model similar? How is it different?

Draw Conclusions

1. What is the minimum amount of information that you need to know in order to identify a neutral atom of an element?
2. If you made models of the isotopes boron-10 and boron-11, how would these models be different?

Compare your models with those of other students. Discuss any differences you find among the models.

"Baring the Atom's Mother Heart"

from Selu: Seeking the Corn-Mother's Wisdom

by Marilou Awiakta

Respond to the Reading

1. What made the atom destructive?
2. How did the author's mother explain the atom to her?
3. Is this a positive or negative explanation of the atom?

Author Marilou Awiakta was raised near Oak Ridge National Laboratory, a nuclear research laboratory in Tennessee where her father worked. She is of Cherokee and Irish descent. This essay resulted from conversations the author had with writer Alice Walker. It details the author's concern with nuclear technology.

"What is the atom, Mother? Will it hurt us?"

I was nine years old. It was December 1945. Four months earlier, in the heat of an August morning—Hiroshima. Destruction. Death. Power beyond belief, released from something invisible. Without knowing its name, I'd already felt the atoms' power in another form…

"What is the atom, Mother? Will it hurt us?"

"It can be used to hurt everybody, Marilou. It killed thousands of people in Hiroshima and Nagasaki. But the atom itself. . . ? It's invisible, the smallest bit of matter. And it's in everything. Your hand, my dress, the milk you're drinking—. . .

. . . Mother already had taught me that beyond surface differences, everything is [connected]. It seemed natural for the atom to be part of this connection. At school, when I was introduced to Einstein's theory of relativity—that energy and matter are one—I accepted the concept easily.

Understanding Literature

Refrain Refrains are emotionally charged words or phrases that are repeated throughout a literary work and can serve a number of purposes. They often are used to emphasize a central idea. In this work, the refrain is when the author asks, "What is the atom, Mother? Will it hurt us?" These words are emotionally charged because they remind the reader of the real-life example of how atoms can be used to hurt people, such as the bombing of Hiroshima and Nagasaki. The nine-year-old girl shows respect for her mother by relying on her wisdom. Her mother responds by explaining that the atom, itself, is not harmful. Do you think the refrain helps the reader understand the importance of the atom?

Science Connection Nuclear fission, or splitting atoms, is the breakdown of the atomic nucleus. It occurs when a particle, such as a neutron, strikes the nucleus of a uranium atom, splitting the nucleus into two fission fragments and two or three individual neutrons. These released neutrons ultimately cause a chain reaction by splitting more nuclei and releasing more neutrons. When it is uncontrolled, this chain reaction results in a devastating explosion.

Linking Science and Writing

Nature Poem Write a short poem about some element you learned about in this chapter. Think of the element as more than just atoms, because there can be power in atoms when they are bonded. Think of the element and how it is used in a practical way. For instance, metals can be used for the manufacturing of various tools and structures.

Career Connection

Chemist

Chemist Ahmed H. Zewail is a professor of chemistry and physics and the director of the Laboratory for Molecular Sciences at the California Institute of Technology. He was awarded the 1999 Nobel Prize in Chemistry for his research. Zewail and his research team use lasers with fast pulses that last only about a quadrillionth of a second to record the making and breaking of chemical bonds. Originally from Egypt, he has been bestowed many honors.

SCIENCE*Online* To learn more about careers in chemistry and physics, visit the Glencoe Science Web site at **science.glencoe.com.**

Chapter 1 Study Guide

Reviewing Main Ideas

Section 1 Why do atoms combine?

1. The electrons in the electron cloud of an atom are arranged in energy levels.
2. Each energy level can hold a specific number of electrons. *Is the outer energy level of the element shown here stable? Explain.*

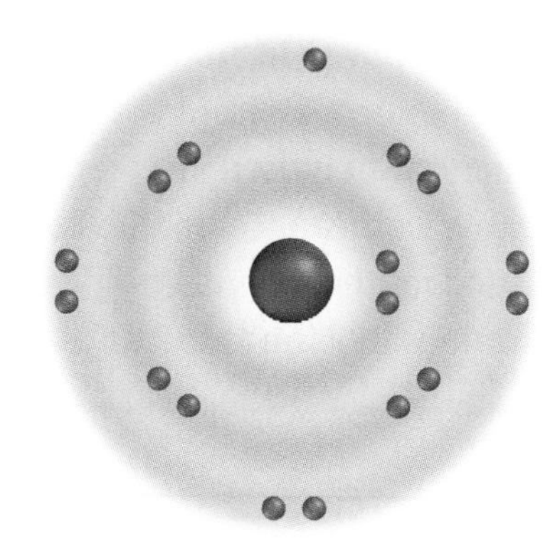

3. The periodic table supplies a great deal of information about the elements, including the numbers of protons and electrons.
4. The number of electrons in the outer energy level of the atom increases across any period of the periodic table.
5. The noble gas elements are stable because their outer energy levels are stable. These elements are called noble because they do not readily form compounds with other elements.
6. Electron dot diagrams show the electrons in the outer energy level of an atom. This tells you how that element behaves. *How do you think the element shown here might react?*

Section 2 How Elements Bond

1. An atom can become stable by gaining, losing, or sharing electrons so that its outer energy level is stable.
2. Ionic bonds form when a metal atom loses one or more electrons and a nonmetal atom gains one or more electrons. *What compound will form from the elements that are shown here?*

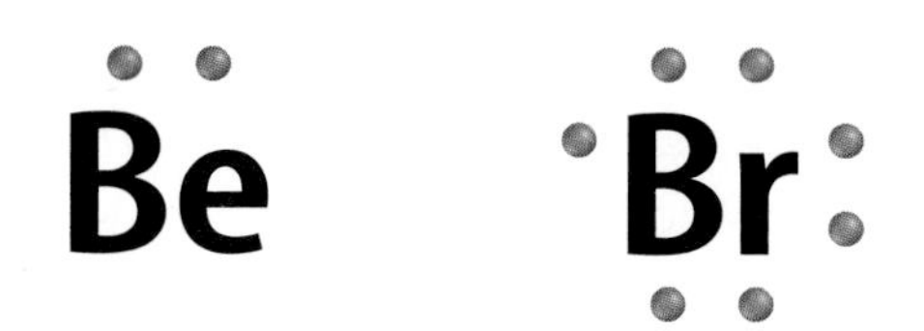

3. Covalent bonds are created when two or more nonmetal atoms share electrons.
4. The unequal sharing of electrons results in a polar covalent bond. In these bonds the electrons tend to spend more time near one side of the molecule than the other.
5. Polar molecules have partial positive areas and partial negative areas. In water, the hydrogens have a partial positive charge and the oxygen has a partial negative charge.
6. A chemical formula indicates which elements and how many atoms of each are present in a compound.

After You Read

Using what you learned in this chapter, explain the difference between polar covalent bonds and covalent bonds on the inside portion of your Foldable.

Visualizing Main Ideas

Complete the following concept map on types of bonds.

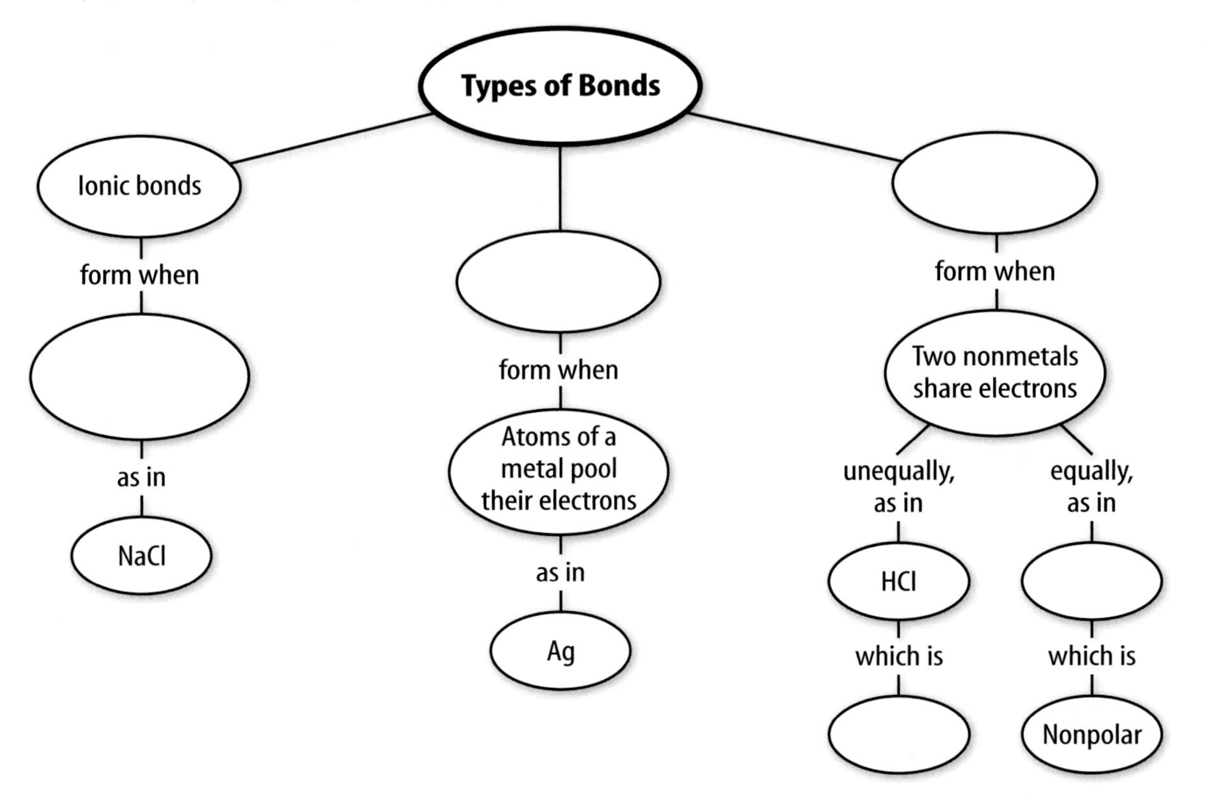

Vocabulary Review

Vocabulary Words

- **a.** chemical bond
- **b.** chemical formula
- **c.** compound
- **d.** covalent bond
- **e.** electron cloud
- **f.** electron dot diagram
- **g.** ion
- **h.** ionic bond
- **i.** metallic bond
- **j.** molecule
- **k.** polar bond

Study Tip

Use repetition to help you memorize facts. For example, when trying to memorize elements and their symbols, write them several times on a piece of paper until you know them.

Using Vocabulary

Distinguish between the terms in each of the following pairs.

1. ion, molecule

2. molecule, compound

3. electron dot diagram, ion

4. chemical formula, molecule

5. ionic bond, covalent bond

6. electron cloud, electron dot diagram

7. covalent bond, polar bond

8. compound, formula

9. metallic bond, ionic bond

Chapter 1 Assessment

Checking Concepts

Choose the word or phrase that best answers the question.

1. Which term represents a compound?
A) equation
B) formula
C) chemical symbol
D) number

2. Which of the following is a covalently bonded molecule?
A) Cl_2
B) air
C) Ne
D) salt

3. Which of the following describes what is represented by the symbol Cl^-?
A) an ionic compound
B) a polar molecule
C) a negative ion
D) a positive ion

4. What happens to electrons in the formation of a polar covalent bond?
A) They are lost.
B) They are gained.
C) They are shared equally.
D) They are shared unequally.

5. Which of the following compounds is unlikely to contain ionic bonds?
A) NaF
B) CO
C) LiCl
D) $MgBr_2$

6. Which term describes the units that make up compounds with covalent bonds?
A) ions
B) molecules
C) salts
D) acids

7. In the chemical formula CO_2, the subscript 2 shows which of the following?
A) There are two oxygen ions.
B) There are two oxygen atoms.
C) There are two CO_2 molecules.
D) There are two CO_2 compounds.

8. Which term describes the units that make up substances formed by ionic bonding?
A) ions
B) molecules
C) acids
D) atoms

9. Which is NOT true about the molecule H_2O?
A) It contains two hydrogen atoms.
B) It contains one oxygen atom.
C) It is a polar covalent compound.
D) It is an ionic compound.

10. What is the number of the group in which the elements have a stable outer energy level?
A) 1
B) 13
C) 16
D) 18

Thinking Critically

11. Groups 1 and 2 form many compounds with Groups 16 and 17. Explain why.

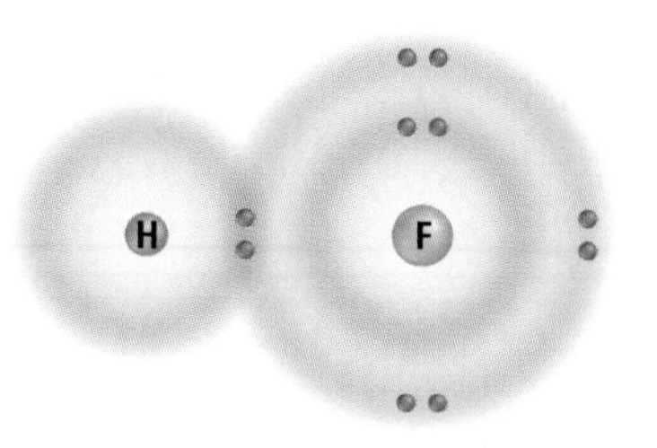

12. What type of bond is shown here? Explain.

13. When salt dissolves in water, the sodium and chloride ions separate. Explain why this might occur.

14. Both cesium, in period 6, and lithium, in period 2, are in the alkali metals family. Cesium is more reactive. Explain this using the energy step diagram in **Figure 4.**

15. Use the fact that water is a polar molecule to explain why water has a much higher boiling point than other molecules of its size.

Developing Skills

16. Predicting If equal masses of CuCl and $CuCl_2$ decompose into their components—copper and chlorine—predict which compound will yield more copper. Explain.

17. **Concept Mapping** Draw a concept map starting with the term chemical bond and use all the vocabulary words.

18. **Recognizing Cause and Effect** A helium atom has only two electrons. Why does helium behave as a noble gas?

19. **Drawing Conclusions** A sample of an element can be drawn easily into wire and conducts electricity well. What kind of bonds can you conclude are present?

20. **Making and Using Tables** Fill in the second column of the table with the number of metal atoms in one unit of the compound. Fill in the third column with the number of atoms of the nonmetal in one unit.

Formulas of Compounds

Compound	No. Metal Atoms	No. Nonmetal Atoms
Cu_2O		
Al_2S_3		
NaF		
$PbCl_4$		

Performance Assessment

21. **Display** Make a display featuring one of the element families described in this chapter. Include electronic structures, electron dot diagrams, and some compounds they form.

TECHNOLOGY

Go to the Glencoe Science Web site at **science.glencoe.com** or use the **Glencoe Science CD-ROM** for additional chapter assessment.

THE PRINCETON REVIEW

Test Practice

A chemist is investigating the structure and properties of some molecules. The table below lists some molecules and their polarities.

Polarities of Molecules

Molecule Name	Chemical Symbol	Molecule Polarity
Carbon dioxide	CO_2	Nonpolar
Chloroform	$CHCl_3$	Polar
Carbon tetrachloride	CCl_4	Nonpolar
Nitrogen trifluoride	NF_3	Polar

Study the chart and answer the following questions.

1. According to the chart, how many carbon-containing molecules are nonpolar?
   A) 1
   B) 0
   C) 2
   D) 3

2. *Like dissolves like* is a common rule that applies to polar and nonpolar substances. It means that two polar substances will mix, but a polar and nonpolar substance will not. According to this information, which of the above molecules will mix with water, a polar substance?
   F) carbon dioxide and chloroform
   G) carbon tetrachloride and carbon dioxide
   H) nitrogen trifluoride and carbon tetrachloride
   J) chloroform and nitrogen trifluoride

CHAPTER 2

Chemical Reactions

Oil refineries like this one provide the starting materials for thousands of chemical reactions. Here, crude oil is separated into many smaller chemical substances. Some of these are used as fuel—they react with oxygen to produce energy for cooking, heating, and transportation. By chemical reactions, others are transformed into plastics, medicines, and fibers for clothing and carpets. In this chapter you will learn about chemical changes and how to describe them using chemical equations. You will see also how the speed of chemical reactions can be controlled.

What do you think?

Science Journal Look at the picture below with a classmate. Discuss what you think this might be. Here's a hint: *It grows but not in a garden.* Write your answer in your Science Journal.

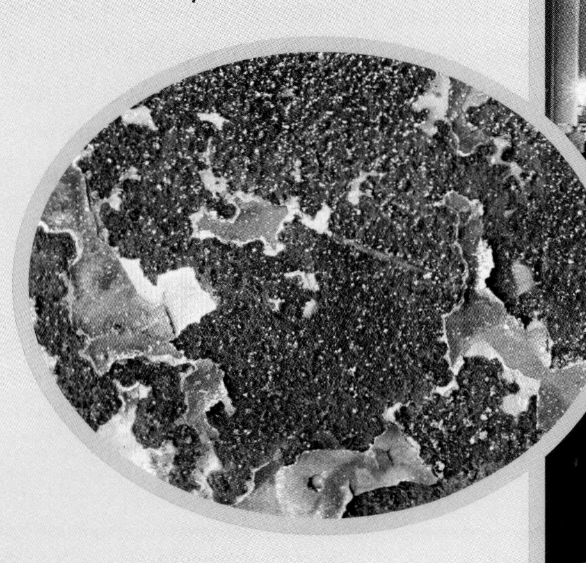

You can see substances changing every day. Fuels burn giving energy to cars and trucks. Green plants convert carbon dioxide and water into oxygen and sugar. Cooking an egg or baking bread causes changes too. These changes are called chemical reactions. In this activity you will observe a common chemical change.

Identify a chemical reaction

WARNING: *Do not touch the test tube. It will be hot. Use extreme caution around an open flame. Point test tubes away from you and others.*

1. Place 3 g of sugar into a large test tube.
2. Carefully light a laboratory burner.
3. Using a test-tube holder, hold the bottom of the test tube just above the flame for 45 s or until something happens.
4. Observe any change that occurs.

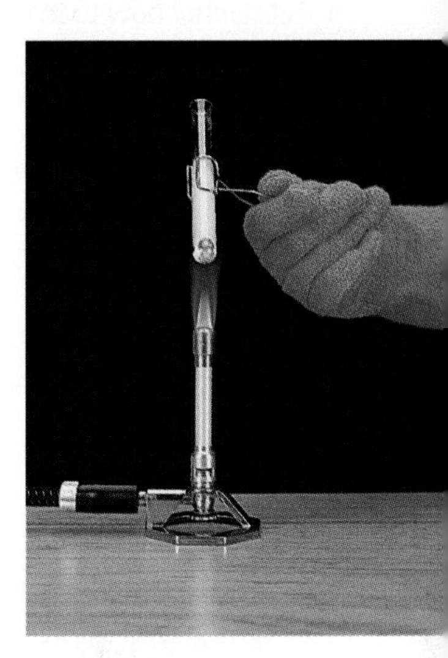

Observe

Describe in your Science Journal the changes that took place in the test tube. What do you think happened to the sugar? Was the substance that remained in the test tube after heating the same as the substance you started with?

Before You Read

FOLDABLES Reading & Study Skills

Making a Question Study Fold **Asking yourself questions while you read helps you stay focused and better understand chemical reactions.**

1. Place a sheet of notebook paper in front of you so the short side is at the top and the holes are on the right side. Fold the paper in half from the left side to the right side.
2. Through the top thickness of paper, cut along every third line from the outside edge to the centerfold, forming tabs as shown.
3. Before you read the chapter, write several questions you have about chemical reactions on the front of the tabs. As you read the chapter, add more questions.

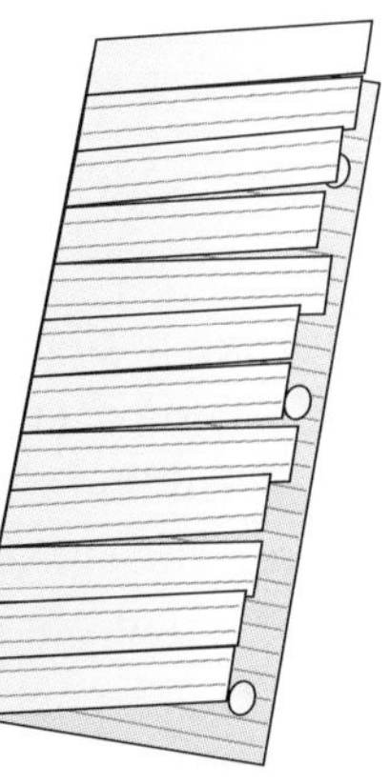

SECTION 1

Chemical Formulas and Equations

As You Read

What You'll Learn

- **Determine** whether or not a chemical reaction is occurring.
- **Determine** how to read and understand a balanced chemical equation.
- **Examine** some reactions that release energy and others that absorb energy.

Vocabulary

chemical reaction
reactant
product
chemical equation
endothermic reaction
exothermic reaction

Why It's Important

Chemical reactions warm your home, digest your food, cook your meals, and power cars and trucks.

Physical or Chemical Change?

You can smell a rotten egg and see the smoke from a campfire. Signs like these tell you that a chemical reaction is taking place. Other evidence might be less obvious, but clues are always present to announce that a reaction is under way.

Matter can undergo two kinds of changes—physical and chemical. Physical changes in a substance affect only physical properties, such as its size and shape, or whether it is a solid, liquid, or gas. For example, when water freezes, its physical state changes from liquid to solid, but it's still water.

In contrast, chemical changes produce new substances that have properties different from those of the original substances. The rust on a bike's handlebars, for example, has properties different from those of the metal around it. A process that produces chemical change is a **chemical reaction.**

To compare physical and chemical changes, look at the newspaper shown in **Figure 1.** If you fold it, you change its size and shape, but it is still newspaper. Folding is a physical change. If you use it to start a fire, it will burn. Burning is a chemical change because new substances result. How can you recognize a chemical change? **Figure 2** shows what to look for.

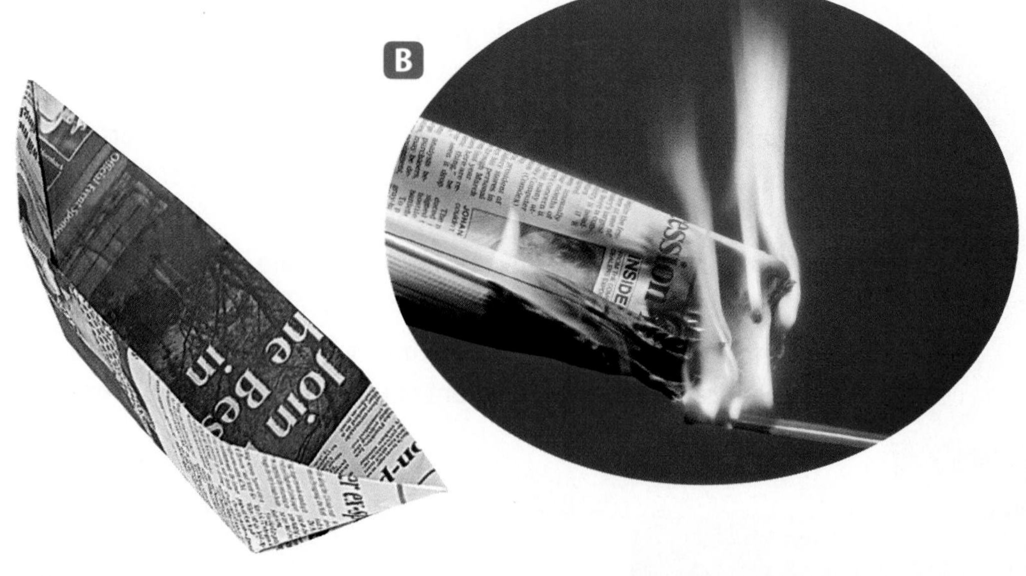

Figure 1
Newspaper can undergo both physical and chemical changes. A Folding changes the shape but not the substance; therefore, it is a physical change. B Burning creates new substances; therefore, burning is a chemical change.

Figure 2

Chemical reactions take place when chemicals combine to form new substances. Your senses—sight, taste, hearing, smell, and touch—can help you detect chemical reactions in your environment.

▼ TASTE A boy grimaces after sipping a milk that has gone sour due to a chemical reaction.

▲ SIGHT When you spot a firefly's bright glow, you are seeing a chemical reaction in progress—two chemicals are combining in the firefly's abdomen and releasing light in the process. The holes in a slice of bread are visible clues that sugar molecules were broken down by yeast cells in a chemical reaction that produces carbon dioxide gas. The gas caused the bread dough to rise.

▲ SMELL AND TOUCH Billowing clouds of acrid smoke and waves of intense heat indicate that chemical reactions are taking place in this

▲ HEARING A Russian cosmonaut hoists a flare into the air after landing in the ocean during a training exercise. The hissing sound of the burning flare is the result of a chemical

Observing a Chemical Change

Procedure

1. Place about 1 g of **baking soda** in an **evaporating dish.** Add 2 mL of **white vinegar.**
2. Allow the mixture to dry.
3. Examine the result and compare it with baking soda. Do they look the same?
4. To find out, add 2 mL of vinegar to the substance remaining in the dish and observe.

Analysis

1. Did a chemical reaction occur in step 1? In step 4? Explain.
2. Are the chemical properties of the residue the same as those of baking soda? Explain.

Chemical Equations

To describe a chemical reaction, you must know which substances react and which substances are formed in the reaction. The substances that react are called the reactants (ree AK tunts). **Reactants** are the substances that exist before the reaction begins. The substances that form as a result of the reaction are called the **products.**

When you mix baking soda and vinegar, a vigorous chemical reaction occurs. The mixture bubbles and foams up inside the container as you can see in **Figure 3.**

Baking soda and vinegar are the common names for the reactants in this reaction, but they also have chemical names. Baking soda is the compound sodium hydrogen carbonate (often called sodium bicarbonate), and vinegar is a solution of acetic (uh SEE tihk) acid in water. These are the reactants. What are the products? You saw bubbles form when the reaction occurred, but is that enough of a description?

Describing What Happens Bubbles tell you that a gas has been produced, but they don't tell you what kind of gas. Are bubbles of gas the only product, or do some atoms from the vinegar and baking soda form something else? What goes on in the chemical reaction can be more than what you see with your eyes. Chemists try to find out which reactants are used and which products are formed in a chemical reaction. Then, they can write it in a shorthand form called a **chemical equation.** This equation tells chemists at a glance the reactants, products, and the proportions of each substance present. This is very important as you will see later.

Reading Check *What does a chemical equation tell chemists?*

Figure 3
The bubbles tell you only that a chemical reaction has taken place but they tell you nothing more about the reaction. *How might you find out whether a new substance has formed?*

Table 1 Reactions Around the Home

Reactants		Products
Baking soda + Vinegar	→	Gas + White solid
Charcoal + Air	→	Ash + Gas + Heat
Iron + Air	→	Rust
Silver + Air	→	Black tarnish
Gas (kitchen range) + Air	→	Gas + Heat
Tea + Lemon juice	→	Lighter-colored tea

Using Words One way you can describe a chemical reaction is with an equation that uses words to name the reactants and products. The reactants are listed on the left side of an arrow, separated from each other by plus signs. The products are placed on the right side of the arrow, also separated by plus signs. The arrow between the reactants and products represents the changes that occur during the chemical reaction. When reading the equation, the arrow is read as *produces.*

You can begin to think of processes as chemical reactions even if you do not know the names of all the substances involved. **Table 1** can help you begin to think like a chemist. It shows the word equations for chemical reactions you might see around your home. See how many other reactions you can find. Look for the signs you have learned that indicate a reaction might be taking place. Then try to write them in the form shown in the table.

If you look carefully, you'll spot many chemical reactions going on right in your own kitchen. To find out more about chemical reactions, see the **Kitchen Chemistry Field Guide** at the back of the book.

Using Chemical Names Many chemicals used around the home have common names. For example acetic acid dissolved in water is called vinegar. Some chemicals, such as baking soda, have two common names—it also is known as sodium bicarbonate. However, chemical names are usually used in word equations instead of common names. In the baking soda and vinegar reaction, you already know the chemical names of the reactants—sodium hydrogen carbonate and acetic acid. The names of the products are sodium acetate, water, and carbon dioxide. The word equation for the reaction is as follows.

Acetic acid + Sodium hydrogen carbonate ⟶
Sodium acetate + Water + Carbon dioxide

Life Science INTEGRATION

A color change usually means a chemical reaction has taken place. In the case of autumn leaves, however, just the opposite is true. The colors are always there, but in spring and summer they are masked by green chlorophyll. When growth slows in autumn, chlorophyll production stops and the red and yellow colors become visible.

Using Formulas The word equation for the reaction of baking soda and vinegar is long. That's why chemists use chemical formulas to represent the chemical names of substances in the equation. You can convert a word equation into a chemical equation by substituting chemical formulas for the chemical names. For example, the chemical equation for the reaction between baking soda and vinegar can be written as follows.

$$HC_2H_3O_2 + NaHCO_3 \rightarrow NaC_2H_3O_2 + H_2O + CO_2$$

$HC_2H_3O_2$	$NaHCO_3$	$NaC_2H_3O_2$	H_2O	CO_2
Acetic acid (vinegar)	Sodium hydrogen carbonate (baking soda)	Sodium acetate	Water	Carbon dioxide

Subscripts When you look at chemical formulas, notice the small numbers written to the right of the atoms. These numbers, called subscripts, tell you the number of atoms of each element in that molecule. For example, the subscript 2 in CO_2 means that each molecule of carbon dioxide has two oxygen atoms. If an atom has no subscript, it means that only one atom of that element is in the molecule, so carbon dioxide has one carbon atom.

Conservation of Mass

What happens to the atoms in the reactants when they are converted into products? Do they disappear? No. According to the law of conservation of mass, the mass of the products of a chemical reaction must be the same as the mass of the reactants in that reaction. This principle was stated first by the French chemist Antoine Lavoisier, who is considered the first modern chemist. Lavoisier used logic and scientific methods to study chemical reactions. He proved by his experiments that nothing is lost or created in chemical reactions.

In a way, he showed that chemical reactions are much like mathematical equations. In math equations, the right and left sides of the equation are numerically equal. Chemical equations are similar, but it is the number and kind of atoms that are equal on the two sides. Every atom that appears on the reactant side of the equation also appears on the product side, as shown in **Figure 4.** Atoms are never lost or created in a chemical reaction; however, they do change partners.

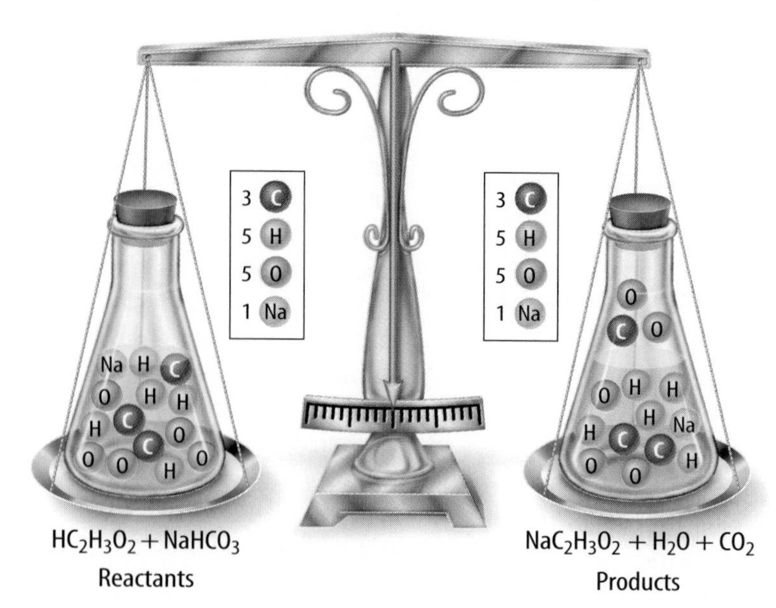

Figure 4
According to the law of conservation of mass, the products of a chemical reaction must contain the same number of atoms of each kind as the reactants.

Figure 5
Keeping silver bright takes frequent polishing, especially in homes heated by gas. Sulfur compounds found in small concentrations in natural gas react with silver, forming black silver sulfide, Ag_2S.

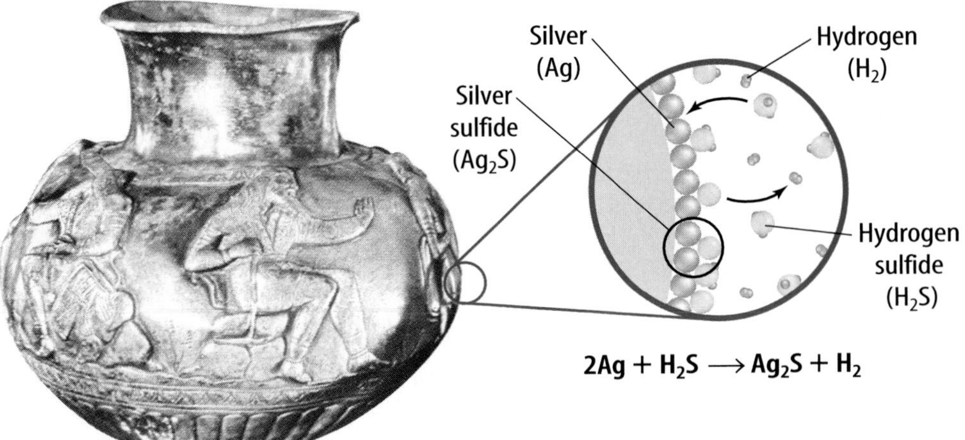

Balancing Chemical Equations

When you write the chemical equation for a reaction, you must observe the law of conservation of mass. Sometimes this is easy, as in the vinegar and baking soda reaction. All you need to do is write the chemical formulas for the reactants and products. Look back at **Figure 4.** It shows that when you count the number of carbon, hydrogen, oxygen, and sodium atoms on each side of the arrow in the equation, you find equal numbers of each kind of atom. This means the equation is balanced.

Reading Check *How do you know if an equation is balanced?*

Not all chemical equations are balanced so easily. For example, silver tarnishes as in **Figure 5** when it reacts with sulfur compounds in the air, such as hydrogen sulfide. The following equation shows what happens when silver tarnishes.

$$Ag + H_2S \rightarrow Ag_2S + H_2$$

Silver, Hydrogen sulfide, Silver sulfide, Hydrogen

Count the Atoms Count the number of atoms of each type in the reactants and in the products. The same numbers of hydrogen and sulfur atoms are on each side, but one silver atom is on the reactant side and two atoms are on the product side. This cannot be true. A chemical reaction cannot create a silver atom, so this equation does not represent the reaction correctly. Place a 2 in front of the reactant Ag and check to see if the equation is balanced now.

To balance chemical equations, place numbers before the formulas as you did just now for Ag. However, you must never change the subscripts written to the right of the atoms in a formula. Changing these numbers changes the identity of the compound.

Research Visit the Glencoe Science Web site at **science.glencoe.com** for more information about balancing chemical equations. Communicate to your class what you learned.

Energy in Chemical Reactions

Often energy is released or absorbed during a chemical reaction. The energy for the welding torch in **Figure 6** is released when hydrogen and oxygen combine to form water.

$$2H_2 + O_2 \rightarrow 2H_2O + \text{energy}$$

Energy Released Where does this energy come from? To answer this question, think about the chemical bonds that break and form when atoms gain, lose, or share electrons. When such a reaction takes place, bonds break in the reactants and new bonds form in the products. In reactions that release energy, the products are more stable, and their bonds have less energy than those of the reactants. The extra energy is released in various forms—light, sound, and heat.

Math Skills Activity

Balancing Equations

Example Problem

Methane and oxygen react to form carbon dioxide, water and heat. You can see how mass is conserved by balancing the equation: $CH_4 + O_2 \rightarrow CO_2 + H_2O$.

Solution

1. *This is what you know:*
   The number of atoms of C, H, and O in reactants and products.
2. *This is what you need to do:*
   Make sure that the reactants and products have equal numbers of atoms of each element. Start with the reactant having the greatest number of atoms.

Reactants	Products	Action
$CH_4 + O_2$	$CO_2 + H_2O$	
have 4 H atoms	need 2 H atoms	Multiply H_2O by 2 to give 4 H atoms
$CH_4 + O_2$	$CO_2 + 2H_2O$	
need 2 more O atoms	have 4 O atoms	Multiply O_2 by 2 to give 4 O atoms

The balanced equation is $CH_4 + 2O_2 \rightarrow CO_2 + 2H_2O$.

Check your answer: Count the carbons, hydrogens, and oxygens on each side.

Practice Problem

Balance the equation $Fe_2O_3 + CO \rightarrow Fe_3O_4 + CO_2$.

For more help, refer to the Math Skill Handbook.

Figure 6
This welding torch burns hydrogen and oxygen to produce temperatures above 3,000°C. It can even be used under water.
What is the product of this chemical reaction?

Energy Absorbed What happens when the reverse is true? In reactions that absorb energy, the reactants are more stable and their bonds have less energy than those of the products.

$$2H_2O + \text{energy} \rightarrow 2H_2 + O_2$$

Water — Hydrogen — Oxygen

In this reaction the extra energy found in the products can be supplied in the form of electricity, as shown in **Figure 7.**

As you have seen, reactions can release or absorb energy of all kinds, including electricity, light, sound, and heat. When heat energy is gained or lost in reactions, special terms are used. **Endothermic** (en duh THUR mihk) **reactions** absorb heat energy. **Exothermic** (ek soh THUR mihk) **reactions** release heat energy. You may notice that the root word *therm* terms refers to heat, as it does in thermos bottles and thermometers.

Heat Released You might already be familiar with several types of reactions that release heat. Burning is an exothermic chemical reaction in which a substance combines with oxygen to produce heat along with light, carbon dioxide, and water.

What type of chemical reaction is burning?

Rapid Release Sometimes energy is released rapidly. For example, charcoal lighter fluid combines with oxygen in the air and produces enough heat to ignite a charcoal fire within a few minutes.

Figure 7
Electrical energy is needed to break water into its components. This is the reverse of the reaction that takes place in the welding torch shown in Figure 6.

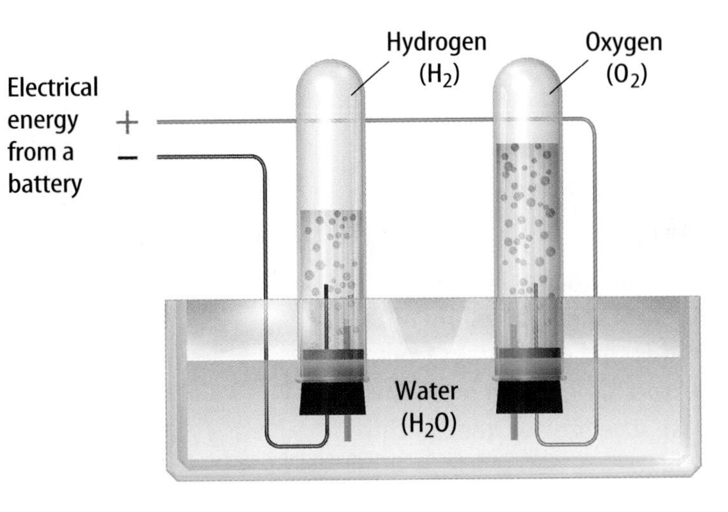

Figure 8
Two exothermic reactions are shown. **A** Lighter fluid combines rapidly with oxygen in air to start a charcoal fire. **B** In a slow reaction, the iron in this wheelbarrow combines with oxygen in air to form rust.

Slow Release Other materials also combine with oxygen but release heat so slowly that you cannot see or feel it happen. This is the case when iron combines with oxygen in the air to form rust. The slow heat release from a reaction also is used in heat packs that can keep your hands warm for several hours. Fast and slow energy release are compared in **Figure 8.**

Heat Absorbed Some chemical reactions and processes need to have heat energy added before they can proceed. An example of an endothermic process that absorbs heat energy is the cold pack shown in **Figure 9.**

The heavy plastic cold pack holds ammonium nitrate and water. The two substances are separated by a plastic divider. When you squeeze the bag, you break the divider so that the ammonium nitrate dissolves in the water. The dissolving process absorbs heat energy which must come from the surrounding environment—the surrounding air or your skin after you place the pack on the injury.

Figure 9
The heat energy needed to dissolve the ammonium nitrate in this cold pack comes from the surroundings—air or, in this case, the bruised knee of this athlete.

Energy in the Equation The word *energy* often is written in equations as either a reactant or a product. Energy written as a reactant helps you think of energy as a necessary ingredient for the reaction to take place. For example, remember the electrical energy needed to break up water into hydrogen and oxygen. It is important to know that energy must be added to make this reaction occur.

Figure 10
You include heat as a product when you write the equation for the reaction shown here.
Why is this so?

Similarly, in the equation for an exothermic reaction, the word *energy* often is written along with the products. This tells you that energy is released. For example, you include energy when writing the reaction that takes place between oxygen and methane in natural gas when you cook on a gas range, as shown in **Figure 10.**

$$CH_4 + 2O_2 \rightarrow CO_2 + 2H_2O + \text{energy}$$

Methane, Oxygen, Carbon dioxide, Water

This heat energy cooks your food.

Although it is not necessary, writing the word *energy* can draw attention to an important aspect of the equation.

Section 1 Assessment

1. Are the following chemical equations balanced? Why or why not?
   a. $Ca + Cl_2 \rightarrow CaCl_2$
   b. $Zn + Ag_2S \rightarrow ZnS + Ag$
   c. $Cl_2 + NaBr \rightarrow NaCl + Br_2$
2. What evidence might tell you that a chemical reaction has occurred?
3. What is the difference between an exothermic and an endothermic reaction?
4. What forms of energy might be needed for a chemical reaction to take place?
5. **Think Critically** After a forest fire, the ashes that are remaining have less mass and take up less space than the trees and vegetation that grew there before the fire. How can this be explained in terms of the law of conservation of mass?

Skill Builder Activities

6. **Recognizing Cause and Effect** After you injure your ankle playing field hockey, you soak your swollen ankle in an Epsom salts solution. Following the directions on the package, you dissolve the salts in water and notice that the water becomes cool. What caused this? **For more help, refer to the** Science Skill Handbook.
7. **Solving One-Step Equations** The equation for the decomposition of silver oxide is $2Ag_2O \rightarrow 4Ag + O_2$. Set up a proportion to calculate the number of silver atoms produced and the number of oxygen molecules released when 1 g of silver oxide is broken down. There are 2.6×10^{21} molecules in 1 g of silver oxide. **For more help, refer to the** Math Skill Handbook.

SECTION 2

Rates of Chemical Reactions

As You Read

***What* You'll Learn**

- **Determine** how to describe and measure the speed of a chemical reaction.
- **Identify** how chemical reactions can be speeded up or slowed down.

Vocabulary

activation energy
rate of reaction
enzyme
inhibitor
catalyst
concentration

***Why* It's Important**

Speeding up useful reactions and slowing down destructive ones can be helpful.

How Fast?

Fireworks explode in rapid succession on a summer night. Old copper pennies darken while they lie forgotten in a drawer. Cooking an egg for two minutes instead of five minutes makes a difference. The amount of time you leave coloring solution on your hair must be timed accurately to give the color you want. Chemical reactions are common in your life. However, notice from these examples that time has something to do with many of them.

As you can see in **Figure 11,** not all chemical reactions take place at the same rate. Some obviously need help to get going, such as fireworks or lighting a campfire. Others seem spontaneous. Does the length of time a chemical reaction occurs affect the end product? Will the stain on your shirt be removed more completely if you soak it longer in bleach? Think about the difference between light toast and burnt toast. Can you do anything to control a chemical reaction?

Figure 11
Reaction speeds vary greatly.

A **Fireworks are over in a few seconds. Metals added to the explosive mixture provide colors—lithium for red, magnesium for white, and copper for blue.**

B **The copper coating on pennies darkens slowly as it reacts with substances it touches. Look at some pennies in your purse or pocket and check their dates to see how long these reactions have been going on.**

Activation Energy

Before a reaction can start, molecules of the reactants have to bump into each other or collide. This makes sense because to form new chemical bonds, atoms have to be close together. But, not just any collision will do. The collision must be strong enough. This means the reactants must smash into each other with a minimum amount of energy. Anything less, and the reaction will not occur. Why is this true?

To form new bonds in the product, old bonds must break in the reactants, and breaking bonds takes energy. To start any chemical reaction, a minimum amount of energy is needed. This energy is called the **activation energy** of the reaction.

Reading Check ***What term describes the minimum amount of energy needed to start a reaction?***

What about reactions that release energy? Is there an activation energy for these reactions too? Yes, even though they release energy later, these reactions need enough energy to start.

One example of a reaction that needs energy to start is the burning of gasoline. You have probably seen movies in which a car plunges over a cliff, lands on the rocks below, and suddenly bursts into flame. But if some gasoline is spilled accidently while filling a gas tank, it probably will evaporate harmlessly in a short time.

Why doesn't this spilled gasoline explode as it does in the movies? The reason is that gasoline needs energy to start burning. That is why there are signs at filling stations like those shown in **Figure 12A** warning you not to smoke. Other signs advise you to turn off ignitions. Why might cars catch fire?

This is similar to the lighting of the Olympic Torch, as shown in **Figure 12B.** Torches designed for each Olympics contain highly flammable materials that cannot be extinguished by high winds or rain. However, they do not ignite until the opening ceremonies when a runner lights the torch using a flame that was kindled in Olympia, Greece, the site of the original Olympic Games.

Research Visit the Glencoe Science Web site at **science.glencoe.com** for more information about fireworks. Communicate to your class what you learned.

Figure 12
Most fuels need energy to ignite.

A Gasoline that is spilled while pumping gas won't burn if safety warnings are observed.

B The Olympic Torch in the 2000 Olympics was lit by athlete Cathy Freeman.

Reaction Rate

Many physical processes are measured in terms of a rate. A rate tells you how much something changes over a given period of time. For example, speed is the rate at which you run or ride your bike. It's the amount of distance you move divided by the time you move. For example, you may jog at a rate of 8 km/h.

Chemical reactions have rates, too. The **rate of reaction** tells us how fast a reaction occurs. To find the rate of a reaction, you can measure either how quickly one of the reactants is disappearing or how quickly one of the products is appearing, as in **Figure 13.** Both measurements tell how the amount of a substance changes per unit of time.

Figure 13
The diminishing amount of wax in this candle as it burns indicates the rate of the reaction.

✔ Reading Check ***What can you measure to determine the rate of a reaction?***

Reaction rate is important in industry usually because the faster the product can be made, the less it costs. However, sometimes fast rates of reaction are undesirable as in the case of reactions that cause food spoilage. In this case, the slower the reaction rate is, the longer the food will stay edible. What conditions control the reaction rate, and how can the rate be changed?

Temperature Changes Rate You can keep the food you buy at the store from spoiling so quickly by putting it in the refrigerator or freezer, as in **Figure 14A.** Food spoiling is a chemical reaction, and the temperature of the food lowers the rate of this reaction. Lowering the temperature slows the reaction.

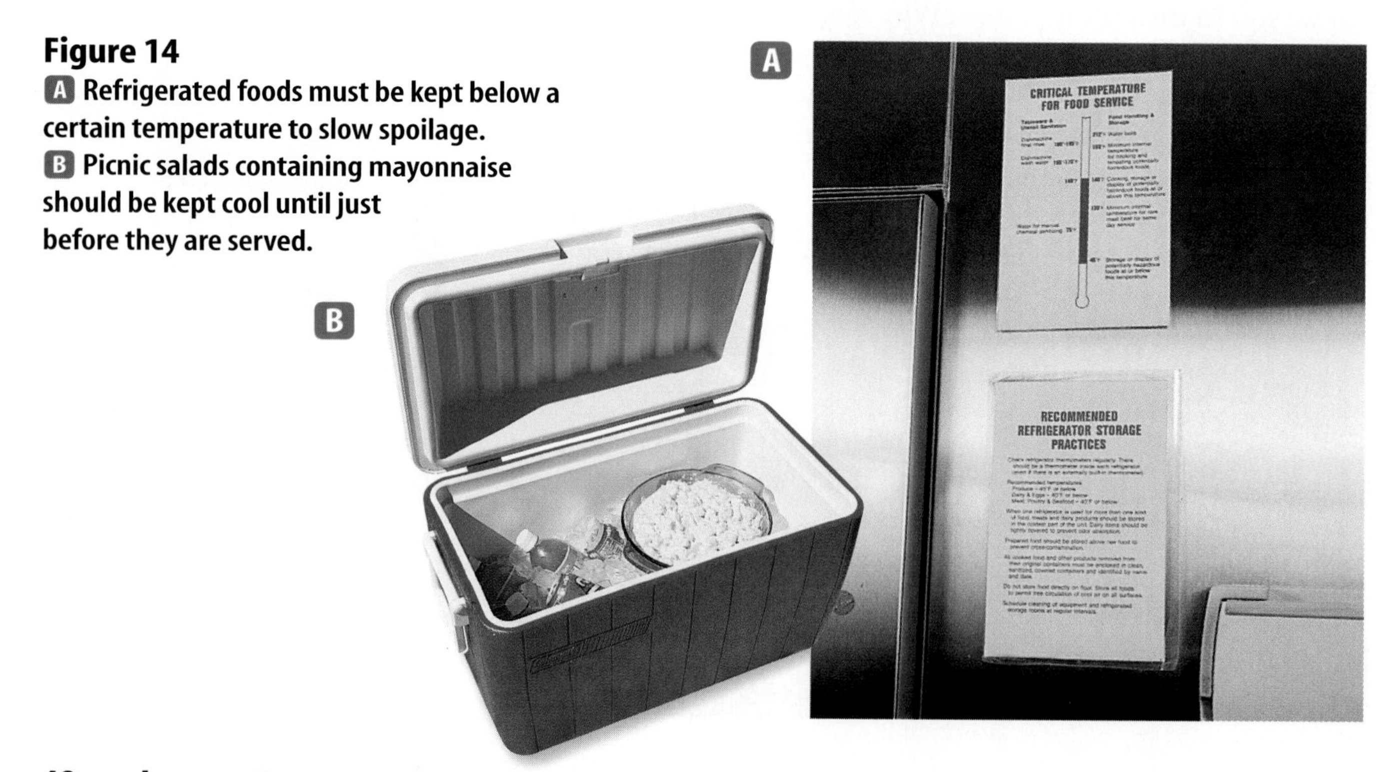

Figure 14
A Refrigerated foods must be kept below a certain temperature to slow spoilage.
B Picnic salads containing mayonnaise should be kept cool until just before they are served.

Meat and fish decompose faster at higher temperatures, producing toxins that can make you sick. Bacteria grow faster at higher temperatures, too, so they reach dangerous levels sooner. Eggs may contain such bacteria, but the chemical reaction that cooks eggs also kills bacteria, so hard-cooked eggs are safer to eat than soft-cooked or raw eggs. All foods should stay chilled until they are served.

Raising Temperature Most chemical reactions speed up when temperature increases. This is because atoms and molecules are always in motion, and they move faster at higher temperatures, as shown in **Figure 15.** Faster molecules collide with each other more often and with greater energy than slower molecules do, so collisions are more likely to provide enough energy to break the old bonds. This is the activation energy.

The high temperature inside an oven speeds up the chemical reactions that turn a liquid cake batter into a more solid, spongy cake. This works the other way, too. Lowering the temperature slows down most reactions. If you set the oven temperature too low accidentally, your cake will not bake properly.

Concentration Affects Rate The more reactant atoms and molecules that are present, the greater the chance is of collisions between them and the faster the reaction rate is. It's like the situation shown in **Figure 16.** When you try to walk through a crowded train station, you're more likely to bump into other people than if the station were not so crowded. The amount of substance present in a certain volume is called the **concentration** of that substance. If you increase the concentration, you increase the number of particles of a substance per unit of volume.

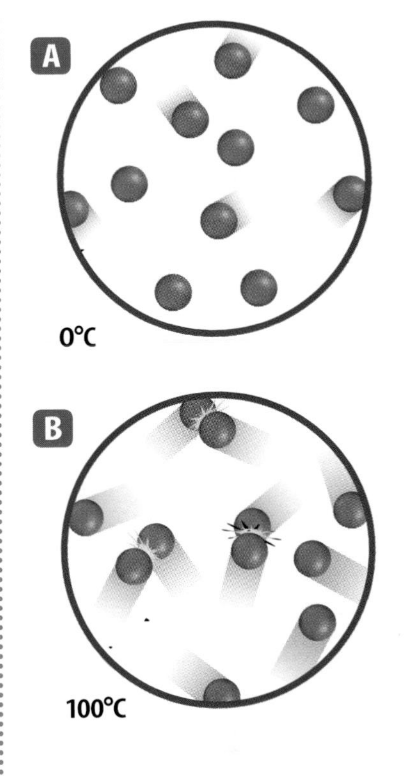

Figure 15
Molecules collide more frequently at high temperatures. This means they are more likely to react. A At 0°C, molecules move slowly and seldom collide. B At 100°C, molcules move rapidly and collide frequently.

A Collisions are more likely in a concentrated solution.

B Fewer collisions occur in a dilute solution.

Figure 16
People are more likely to collide in crowds. Molecules behave similarly.

Figure 17
A **Molecules trapped inside a particle cannot react, so the reaction is slow.** **B** **Breaking up a large particle exposes many more molecules at the surface, so the reaction speeds up.**

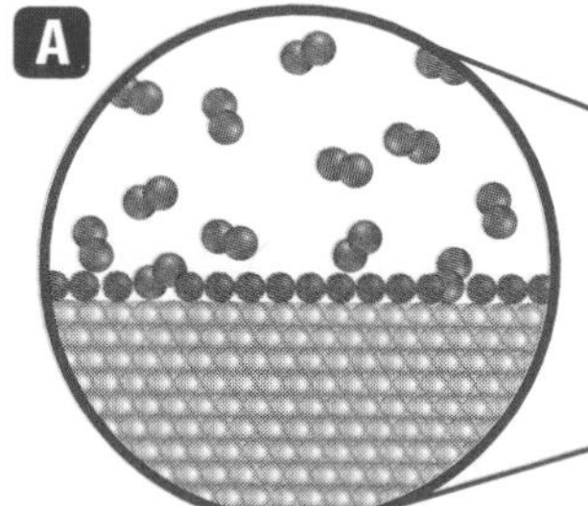

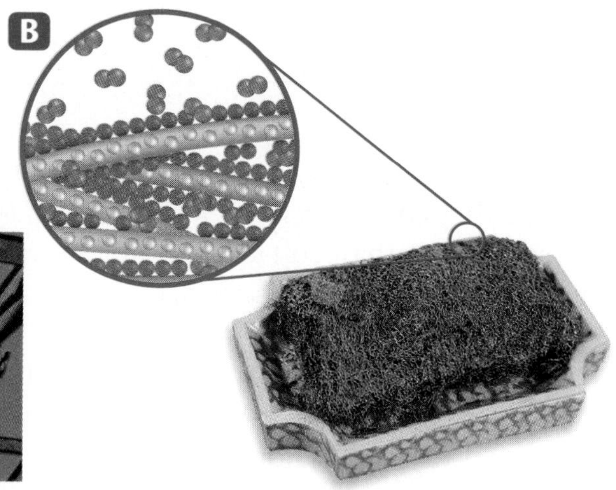

Particle Size Can Change the Rate The size of the reactant particles also affects how fast the reaction can occur. You can easily start a campfire with small twigs, but starting a fire with only large logs would probably not work.

Only the atoms or molecules in the outer layer of the reactant material can touch the other reactants and react. **Figure 17A** shows that when particles are large, most of the atoms are stuck inside and can't react. If the particles are small as in **Figure 17B,** more of the reactant atoms are at the surface and can react.

Identifying Inhibitors

Procedure

1. Look at the ingredients listed on **packages of cereals** and **crackers** in your kitchen.
2. Note the preservatives listed. These are chemical inhibitors.
3. Compare the date on the box with the approximate date the box was purchased to estimate shelf life.

Analysis

1. What is the average shelf life of these products?
2. Why is increased shelf life of such products important?

Slowing Down Reactions

Sometimes reactions occur too quickly. For example, food and medications can undergo chemical reactions that cause them to spoil or lose their effectiveness too rapidly. Luckily, these reactions can be slowed down.

A substance that slows down a chemical reaction is called an inhibitor. An **inhibitor** doesn't completely stop a reaction, but it makes the formation of a certain amount of product take longer, as shown in **Figure 18.** Many cereal boxes contain the compound butyl hydroxytoluene, BHT. The BHT in the packaging material slows the spoiling of the cereal and increases its shelf life.

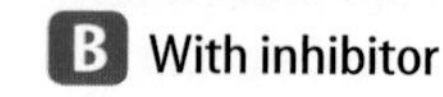

Figure 18
A reaction proceeds for 1 h. **A** **The reaction without an inhibitor is complete.** **B** **The reaction with an inhibitor is not yet complete and is not finished producing product.**

Speeding Up Reactions

Is it possible to speed up a chemical reaction? Yes, you can add a catalyst (KAT uh lihst). A **catalyst** is a substance that speeds up a chemical reaction. Catalysts do not appear in the chemical equation because they are not changed permanently or used up. A reaction using a catalyst will not produce more product than a reaction without a catalyst, but it will produce the same amount of product faster.

Reading Check *What does a catalyst do in a chemical reaction?*

How does a catalyst work? Many catalysts speed up reaction rates by providing a surface for the reaction to take place. Sometimes the reacting molecules are held in a particular position that favors reaction. By aiding the molecules in this way, the catalyst reduces the activation energy needed to start the reaction. When the activation energy is reduced, the reaction rate increases.

Life Science INTEGRATION

When a piece of peeled, raw potato darkens in air, one compound in the potato has combined with oxygen to form a dark substance called melanin. This reaction is catalyzed by an enzyme. Similar reactions occur on the cut surfaces of fruits, such as apples. Notice this darkening the next time you peel potatoes or apples.

Catalytic Converters Catalysts are used in the exhaust systems of cars and trucks to aid fuel combustion. The exhaust passes through the catalyst, which often is in the form of beads coated with metals such as platinum or rhodium. This large surface area speeds the reactions that change incompletely burned substances that are harmful, such as carbon monoxide, into harmless substances, such as carbon dioxide. Similarly, hydrocarbons are changed into carbon dioxide and water. The result of these reactions is cleaner air. These reactions are shown in **Figure 19.**

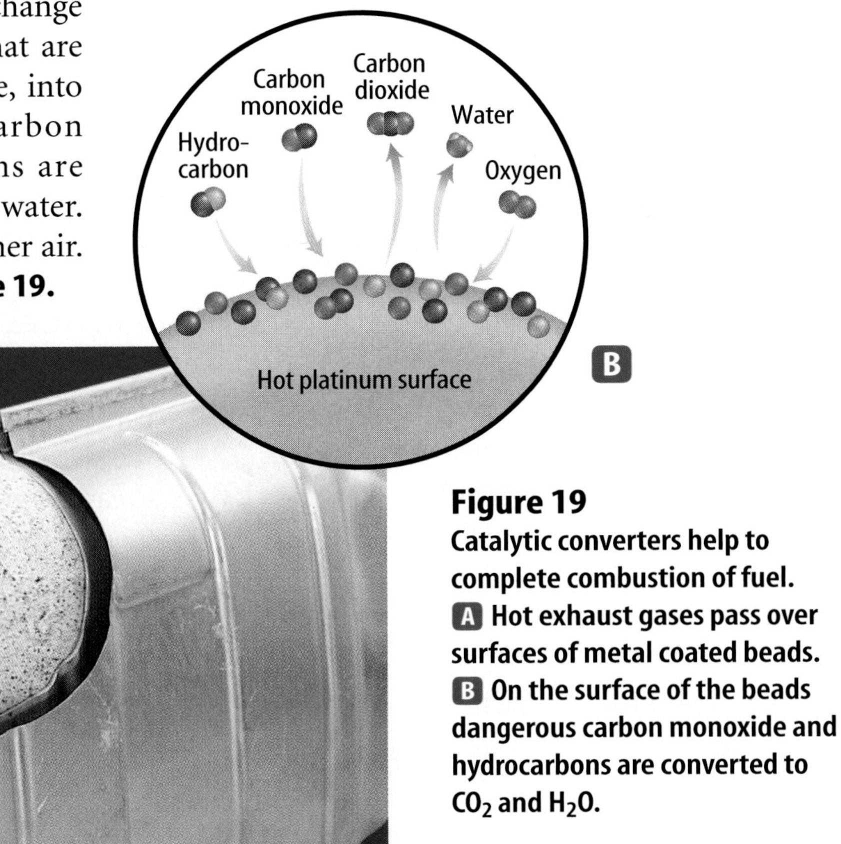

Figure 19
Catalytic converters help to complete combustion of fuel.
A Hot exhaust gases pass over surfaces of metal coated beads.
B On the surface of the beads dangerous carbon monoxide and hydrocarbons are converted to CO_2 and H_2O.

Figure 20
The enzymes in meat tenderizer break down protein in meat, making it more tender.

Enzymes Are Specialists Some of the most effective catalysts are at work in thousands of reactions that take place in your body. These catalysts, called **enzymes** are large protein molecules that speed up reactions needed for your cells to work properly. They help your body convert food to fuel, build bone and muscle tissue, convert extra energy to fat and even to produce other enzymes.

These are complex reactions. Without enzymes, they would occur at rates that are too slow to be useful or they would not occur at all. Enzymes make it possible for your body to function. Like other catalysts, enzymes function by positioning the reacting molecules so that their structures fit together properly. Enzymes are a kind of chemical specialist—an enzyme exists to carry out each type of reaction in your body.

Other Uses Enzymes work outside your body, too. One class of enzymes, called proteases (PROH tee ays es), specializes in protein reactions. They work within cells to break down large, complex molecules called proteins. The meat tenderizer shown in **Figure 20** contains proteases that break down protein in meat, making it more tender. Contact lens cleaning solutions also contain proteases that break down proteins from your eyes that can collect on your lenses and cloud your view.

Section 2 Assessment

1. How can you measure reaction rates?
2. For the general reaction A + B + energy $\longrightarrow$ C, how will the following affect the reaction rate?
   a. increasing the temperature
   b. decreasing the concentration
   c. decreasing particle size
3. Describe how catalysts work to speed up chemical reactions.
4. Explain what is meant by activation energy.
5. **Think Critically** Explain why a jar of spaghetti sauce can be stored for weeks on the shelf in the market but must be placed in the refrigerator after it is opened.

Skill Builder Activities

6. **Predicting** Temperature, concentration, and particle size affect reaction rates. Based on what you know about these factors, how would stirring a reaction mixture affect the reaction rate? **For more help, refer to the** Science Skill Handbook.
7. **Solving One-Step Equations** A chemical reaction is proceeding at a rate of 2 g of product per 45 s. How long will it take to obtain 50 g of product? If adding a catalyst doubles this rate, how long will it take using a catalyst? How long will it take if an inhibiter that halves the rate is added? **For more help, refer to the** Math Skill Handbook.

Activity

Comparing Metals

How can you compare the chemical activity of two substances? One way to do this is to allow each substance to react with a third substance and to compare the results. How might you compare different metals? What third substance would react with the metals to allow comparisons?

What You'll Investigate

Compare the chemical activity of three metals.

Materials

Large test tubes(3)
Test-tube rack
0.1 M hydrochloric acid, HCl (9 mL)
10-mL graduated cylinder
Tongs
Small pieces of zinc, aluminum and copper.

Goals

- **Observe** the chemical reactions between metals and a dilute acid.
- **Compare** and contrast the degree of chemical reactivity among these metals.

Safety Precautions

Handle hydrochloric acid with care. Report any spills immediately. Use only pieces of metal that are less than the size of a pea.

Procedure

1. Put about 3 mL of hydrochloric acid into each of the test tubes in the test tube rack.
2. Use the tongs to drop carefully a piece of one metal into a test tube.
3. Observe carefully what happens and record it in your Science Journal.
4. Repeats steps 1–3 for the two other metals.

Conclude and Apply

1. **Compare** the results of each metal's reaction with the acid.
2. **Sequence** the metals tested in the order of most reactive to least reactive.
3. Your teacher will test the reactive metals again using a lighted wooden splint. Based on these tests, what do you think was one product formed during these reactions?
4. Write a balanced chemical equation for at least one of these reactions.

Compare your results with those of other students in your class. Discuss how any observed differences might be explained.

Activity Design Your Own Experiment

Exothermic or Endothermic?

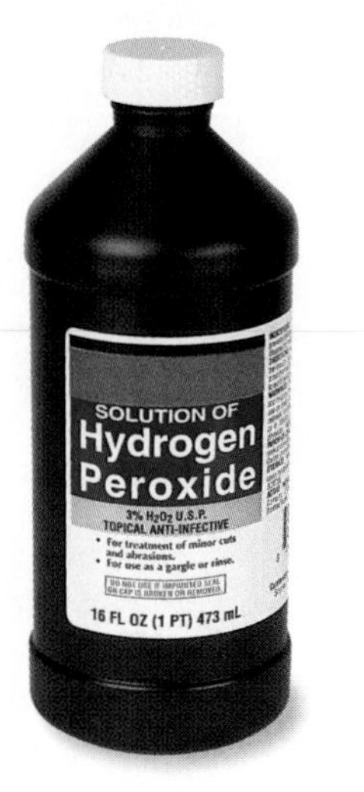

Energy is always a part of a chemical reaction. Some reactions need energy to start. Other reactions release energy into the environment.

Recognize the Problem

How can you tell whether a reaction is exothermic or endothermic?

Form a Hypothesis

What evidence can you find to show that a reaction between hydrogen peroxide and liver or potato is exothermic or endothermic? Think about the difference between these two types of reactions. Make a hypothesis that describes how you can use the reactions between hydrogen peroxide and liver or potato to determine whether a reaction is exothermic or endothermic.

Goals

- **Design** an experiment to test whether a reaction is exothermic or endothermic.
- **Measure** the temperature change caused by a chemical reaction.

Possible Materials

test tubes (8)
test-tube rack
3% hydrogen peroxide solution
raw liver
raw potato
thermometer
stopwatch
clock with second hand
25-mL graduated cylinder

Safety Precautions

Be careful when handling glass thermometers. Test tubes containing hydrogen peroxide should be placed and kept in racks. Dispose of materials as directed by your teacher. Wash your hands when you complete this activity. **WARNING:** *Hydrogen peroxide can irritate skin and eyes and damage clothing.*

Test Your Hypothesis

Plan

1. As a group, look at the list of materials. Decide which procedure you will use to test your hypothesis, and which measurements you will make.
2. **Decide** how you will detect the heat released to the environment during the reaction. Determine how many measurements you will need to make during a reaction.
3. You will get more accurate data if you repeat each experiment several times. Each repeated experiment is called a trial. Use the average of all the trials as your data for supporting your hypothesis.
4. **Decide** what the variables are and what your control will be.
5. Copy the data table in your Science Journal before you begin to carry out your experiment.

Temperature After Adding Liver/Potato

Trial	Temperature After Adding Liver (°C)		Temperature After Adding Potato (°C)	
	Starting	After ___ min	Starting	After ___ min
1				
2				
3				
4				

Do

1. Make sure your teacher approves your plan before you start.
2. Carry out your plan.
3. **Record** your measurements immediately in your data table.
4. **Calculate** the averages of your trial results and record them in your Science Journal.

Analyze Your Data

1. Can you infer that a chemical reaction took place? What evidence did you observe to support this?
2. **Identify** what the variables were in this experiment.
3. **Identify** the control.

Draw Conclusions

1. Do your observations allow you to distinguish between an exothermic reaction and an endothermic reaction? Use your data to explain your answer.
2. Where do you think that the energy involved in this experiment came from? Explain your answer.

Communicating Your Data

Compare the results obtained by your group with those obtained by other groups. Are there differences? **Explain** how these might have occurred.

Synthetic

Diamonds are the most dazzling, most dramatic, most valuable natural objects on Earth. Strangely, these beautiful objects are made of carbon, the same material graphite—the stuff found in pencils—is made of. So why is a diamond hard and clear and graphite soft and black? A diamond's hardness is a result of how strongly its atoms are linked. What makes a diamond transparent is the way its crystals are arranged. The carbon in a diamond is almost completely pure, but there are trace amounts of boron and nitrogen in it. These elements account for the many shades of color found in diamonds—from red to blue to black.

These beauties are also tough. A diamond is the hardest naturally occurring substance on Earth. It's so hard, only a diamond can scratch another diamond. Diamonds are impervious to heat and household chemicals. Their crystal structure allows them to be split (or crushed) along particular lines. Expert diamond cutters know where these fault lines are and can cut diamonds to create gems with glittering sides or facets.

Natural Diamond

A diamond's value isn't based on its beauty and usefulness. It has to do with how it's created. Genuine diamonds are made when carbon is squeezed at high pressures and temperatures in Earth's upper mantle, about 150 km beneath the surface. At that depth, the temperature is about 1,400°C, and the pressure is about 55,000 atmospheres greater than the pressure at sea level.

A researcher at the University of Florida examines a high-pressure, high-temperature machine used to make gem-quality diamonds.

ALMOST THE REAL THING

Diamonds

That's equal in pressure to the 7,000 metric-ton Eiffel Tower sitting on a 127 cm^2 plate! Eventually, most diamonds are carried by magma to the surface of the Earth during volcanic activity.

The conditions that form diamonds and take them to the Earth's crust are relatively rare. That's why the world's supply of natural diamonds is limited. For this reason, diamonds are very valuable—and very expensive. Not surprisingly, people have been eager to make them artificially. As far back as the 1850s, scientists tried to convert graphite into diamonds. It wasn't until 1954 that researchers produced the first synthetic diamonds by compressing carbon under extremely high pressure and heat. Scientists converted graphite powder into tiny diamond crystals using pressure of more than one million pounds per square inch, and at a temperature of about 1,700°C for about 16 hours. In 1961, a technique that used a shock wave produced diamond powder. In 1970, researchers first created gem-quality synthetic diamonds.

Synthetic Diamond

Synthetic diamonds are human-made, but they're not fake. They have all the properties of natural diamonds, from hardness to excellent heat conductivity. Experts claim to be able to detect synthetics because they contain tiny amounts of metal (used in their manufacturing process) and have a different luminescence than natural diamonds.

Though synthetics are virtually identical to natural diamonds, most people don't want to buy them as jewels, because they think of them as imitations. In fact, most synthetics are made for industrial use. One major reason is that making small synthetic diamonds is cheaper than finding small natural ones. The other reason is that synthetics can be made to a required size and shape. Still, if new techniques bring down the cost of producing large, gem-quality synthetic diamonds, they may one day compete with natural diamonds as jewelry.

CONNECTIONS Research Investigate the history of diamonds—natural and synthetic. Explain the differences between them and their uses. Share your findings with the class.

SCIENCE *Online*

For more information, visit science.glencoe.com

Chapter 2 Study Guide

Reviewing Main Ideas

Section 1 Describing a Chemical Reaction

1. Chemical reactions often cause observable changes, such as a change in color or odor, a release or absorbtion of heat or light, or a release of gas. *What clue indicates that a chemical reaction is taking place here?*

2. A chemical equation is a shorthand method of writing what happens in a chemical reaction. Chemical equations use symbols to represent the reactants and products of a reaction, and sometimes show whether energy is produced or absorbed. *Should energy be included in the reaction shown here? Explain your answer.*

3. The law of conservation of mass requires that the same number of atoms of each element be in the products as in the reactants of a chemical equation. This is true in every balanced chemical equation.

Section 2 Rates of Chemical Reactions

1. The rate of reaction is a measure of how quickly a reaction occurs.

2. All reactions have an activation energy—a minimum amount of energy required to start the reaction. Reactions with low activation energies occur rapidly. Those with high activation energies occur slowly or perhaps not at all.

3. The rate of a chemical reaction can be influenced by the temperature, the concentration of the reactants, and the size of the reactant particles.

4. Catalysts can speed up a reaction without being used up. Inhibitors slow down the rate of reaction. *What is the role of BHT in these cereal boxes?*

5. Many catalysts speed reactions by offering a large surface area on which reactions can take place. Others hold molecules in positions that are favourable for reaction.

6. Enzymes are large protein molecules that act as catalysts in many reactions that take place in your body cells.

After You Read

Under the tabs of your Foldable, write answers to the questions you recorded on the tabs at the beginning of the chapter.

Visualizing Main Ideas

Complete the following concept map on chemical reactions.

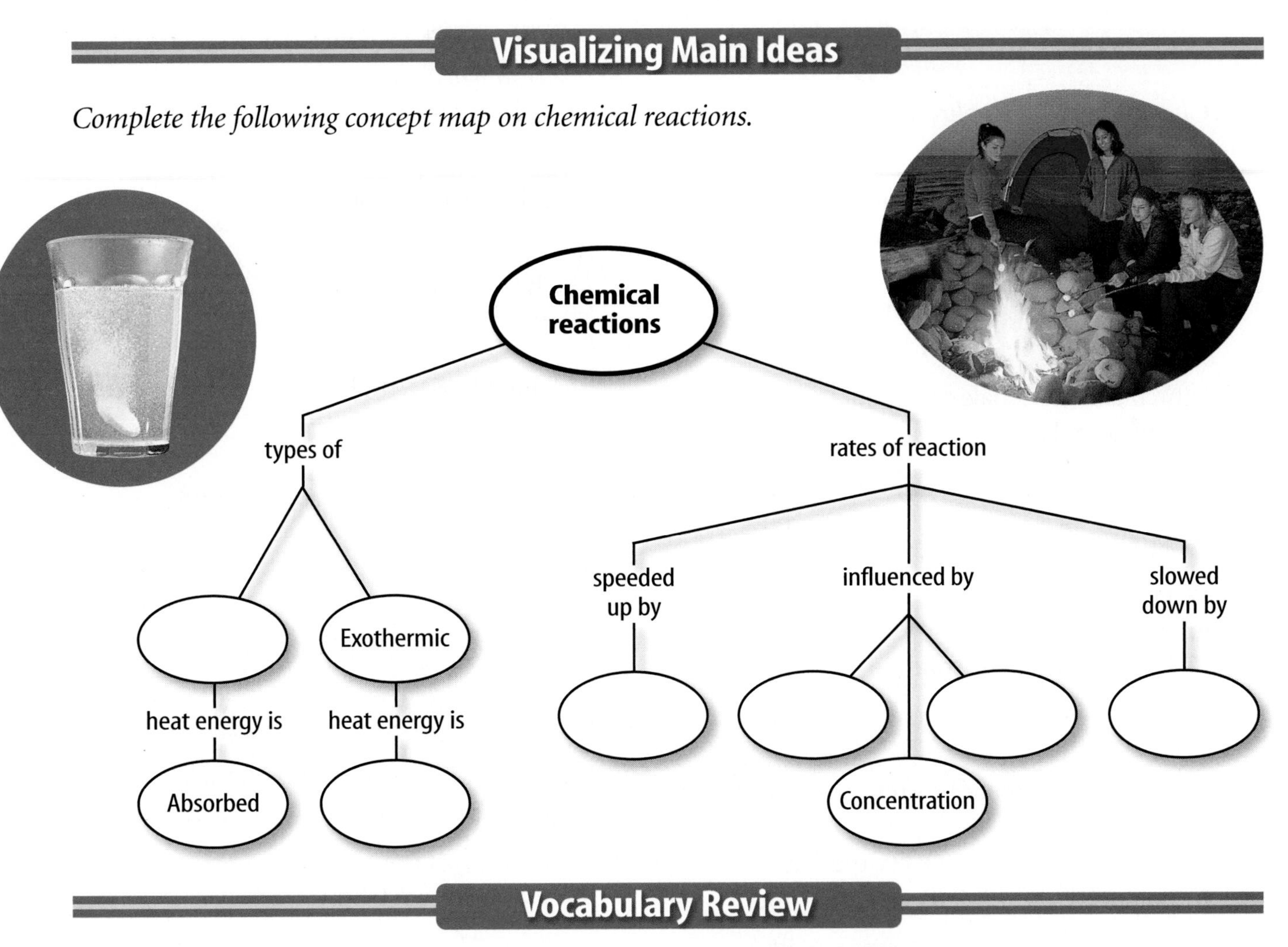

Vocabulary Review

Vocabulary Words

a. activation energy
b. catalyst
c. chemical reaction
d. concentration
e. chemical equation
f. endothermic reaction
g. enzyme
h. exothermic reaction
i. inhibitor
j. product
k. rate of reaction
i. reactant

Study Tip

Keep all your homework assignments, and read them over. Be sure that you understand any questions you answered incorrectly.

Using Vocabulary

Explain the differences between the vocabulary terms in each of the following sets.

1. exothermic reaction, endothermic reaction
2. activation energy, rate of reaction
3. reactant, product
4. catalyst, inhibitor
5. concentration, rate of reaction
6. chemical equation, reactant
7. inhibitor, product
8. catalyst, chemical equation
9. rate of reaction, enzyme

Chapter 2 Assessment

Checking Concepts

Choose the word or phrase that best answers the question.

1. A balanced chemical equation must have the same number of atoms of each of these on both sides.
   A) atoms **C)** molecules
   B) elements **D)** compounds

2. Which is NOT a balanced equation?
   A) $CuCl_2 + H_2S \longrightarrow CuS + 2HCl$
   B) $AgNO_3 + NaI \longrightarrow AgI + NaNO_3$
   C) $2C_2H_6 + 7O_2 \longrightarrow 4CO_2 + 6H_2O$
   D) $MgO + Fe \longrightarrow Fe_2O_3 + Mg$

3. Which of these is a chemical change?
   A) Paper is shredded.
   B) Liquid wax turns solid.
   C) A raw egg is broken.
   D) Soap scum forms.

4. Which of these reactions releases heat energy?
   A) unbalanced **C)** exothermic
   B) balanced **D)** endothermic

5. Which statement about the law of conservation of mass is false?
   A) The mass of reactants must equal the mass of products.
   B) All the atoms on the reactant side of an equation are also on the product side.
   C) The reaction creates new types of atoms.
   D) Atoms are not lost, but are rearranged.

6. What does NOT affect reaction rate?
   A) balancing **C)** particle size
   B) temperature **D)** concentration

7. To slow down a chemical reaction, what should you add?
   A) catalyst **C)** inhibitor
   B) reactant **D)** enzyme

8. Which is NOT evidence that a chemical reaction has occurred?
   A) Milk tastes sour.
   B) Steam condenses on a cold window.
   C) A strong odor comes from a broken egg.
   D) A slice of raw potato darkens.

9. Which of the following would decrease the rate of a chemical reaction?
   A) increase the temperature
   B) reduce the concentration of a reactant
   C) increase the concentration of a reactant
   D) add a catalyst

10. Which of these describes a catalyst?
    A) It is a reactant.
    B) It speeds up a reaction.
    C) It appears in the chemical equation.
    D) It can be used in place of an inhibitor.

Thinking Critically

11. Pickled cucumbers remain edible much longer than fresh cucumbers do. Explain.

12. What can you infer about the reaction from the graph shown below?

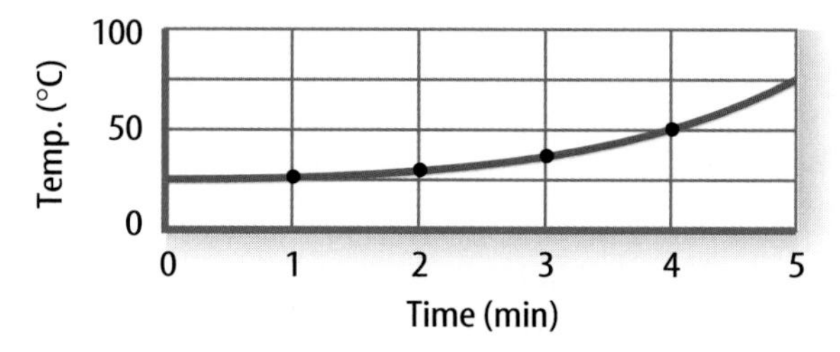

13. A beaker of water in sunlight becomes warm. Has a chemical reaction occurred? Explain.

14. Is $2Ag + S$ the same as Ag_2S? Explain.

15. Apple slices can be kept from browning by brushing them with lemon juice. Infer what role lemon juice plays in this case.

16. Chili can be made using ground meat or chunks of meat. Which would you choose, if you were in a hurry? Explain.

Developing Skills

17. Interpreting Scientific Illustrations The two curves on the graph represent the concentrations of compounds A (blue) and B (red) during a chemical reaction.

a. Which compound is a reactant?

b. Which compound is a product?

c. During which time period is the concentration of the reactant changing most rapidly?

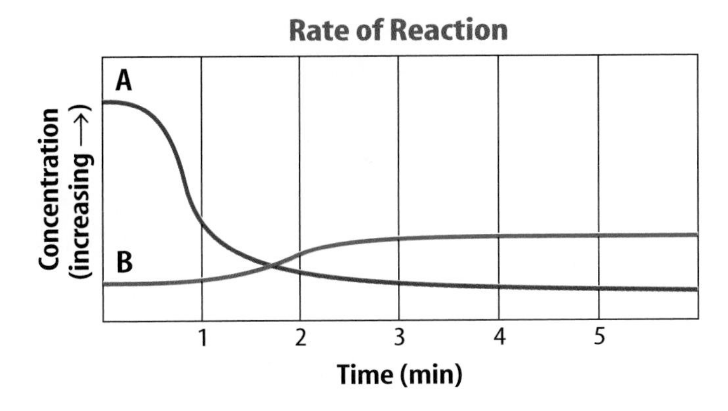

18. Forming Hypotheses You are cleaning out a cabinet beneath the kitchen sink and find an unused steel wool scrub pad that has rusted completely. Will the remains of this pad weigh more or less than when it was new? Explain.

Performance Assessment

19. Poster Make a list of the preservatives in the food you eat in one day. Present your findings to your class in a poster.

Technology

Go to the Glencoe Science Web site at **science.glencoe.com** or use the **Glencoe Science CD-ROM** for additional chapter assessment.

THE PRINCETON REVIEW Test Practice

Mrs. O'Brian's chemistry class is going to carry out a chemical reaction. Below is the reaction written as a chemical equation.

$2H_2O_2$	→	$2H_2O$	+	O_2
Hydrogen peroxide		**Water**		**Oxygen**

Analyze the chemical equation and answer the questions below.

1. According to the equation, which statement best describes what has occurred?

A) Water breaks down to form oxygen and hydrogen peroxide.

B) Oxygen is formed from water and hydrogen peroxide.

C) Hydrogen peroxide breaks down into water and oxygen.

D) Hydrogen peroxide mixes with water to form oxygen.

2. All of the following compounds are involved in the chemical equation EXCEPT ______.

F) water

G) oxygen

H) hydrochloric acid

J) hydrogen peroxide

3. Which of the following does NOT occur in the reaction?

A) New substances are formed.

B) Two molecules of hydrogen peroxide form two molecules of water and one molecule of oxygen.

C) Atoms of oxygen and hydrogen are created.

D) Two substances form from one substance.

CHAPTER 3

Substances, Mixtures, and Solubility

It's halftime, and the band plays. Just as the mixing of notes produces music, the mixing of solids, liquids, and gases produces many of the things around you. From the brass in tubas to the lemonade you drink, you live in a world of mixtures. In this chapter, you'll learn about different kinds of mixtures. You'll learn why some substances form mixtures and others do not, and you'll learn many uses for mixtures.

What do you think?

Science Journal Look at the picture below with a classmate. Discuss what you think is happening. Here's a hint: *Many musicians depend on this solution.* Write your answer or best guess in your Science Journal.

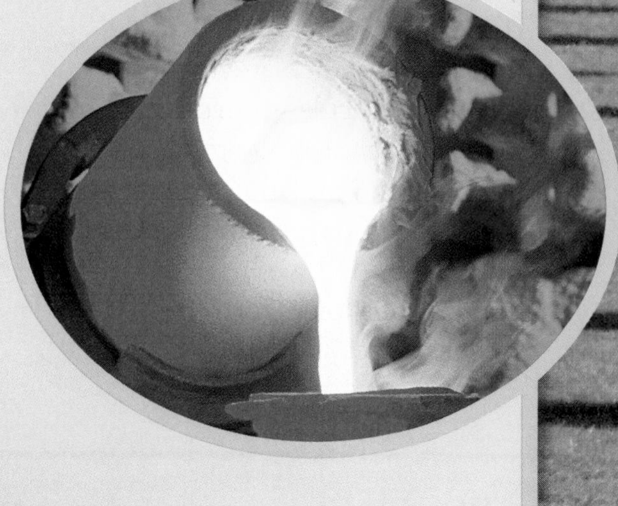

Why do drink mixes come in powder form? Wouldn't it be less costly to sell chunks of drink mix instead of going to the expense of grinding it into powder? Powdered drink mix dissolves faster in water than chunks do because it is divided into smaller particles. More of the mix is exposed to the water. See for yourself how particle size affects the rate that a substance dissolves.

Test the dissolving rate of different-sized particles

1. Pour 400 mL of water into each of the two 600-mL beakers.
2. Carefully grind a bouillon cube into powder using a mortar and pestle.
3. Place the bouillon powder into one beaker and drop a whole bouillon cube into the second beaker.
4. Stir the water in each beaker for 10 s and observe.

Observe

Write a paragraph in your Science Journal comparing the color of the two liquids and the amount of undissolved bouillon at the bottom of each beaker.

Before You Read

Making a Classify Study Fold **Make the following Foldable to help you organize solutions into groups based on their common features.**

1. Place a sheet of paper in front of you so the short side is at the top. Fold the paper in half from the left side to the right side.
2. Fold in the top and the bottom, forming three equal sections. Unfold.
3. Through the top thickness of paper, cut along each of the fold lines to the left fold, forming three tabs. Label the tabs *Solid Solutions, Liquid Solutions,* and *Gaseous Solutions.*
4. As you read the chapter, sort solutions based on their state and list them under the appropriate tabs.

SECTION

What is a solution?

As You Read

What You'll Learn

- **Distinguish** between substances and mixtures.
- **Describe** two different types of mixtures.
- **Explain** how solutions form.
- **Describe** how solids, liquids, and gases can form different types of solutions.

Vocabulary

substance
solution
solute
solvent
precipitate

Why It's Important

The air you breathe, the water you drink, and even parts of your body are all solutions.

Substances

Water, salt water, and pulpy orange juice have some obvious differences. These differences can be explained by chemistry. Think about pure water. No matter what you do to it physically—freeze it, boil it, stir it, or strain it—it still is water. On the other hand, if you boil salt water, the water turns to gas and leaves salt behind. If you strain pulpy orange juice, it loses its pulp. How does chemistry explain these differences? The answer has to do with the chemical compositions of the materials.

Elements Recall that elements consist of atoms. Atoms are the basic building blocks of matter and have unique chemical and physical properties. An atom's identity is determined by the number of protons it has. For example, all atoms that have eight protons are oxygen atoms. The number of protons in an element is fixed—it cannot change unless the element also changes. An element is an example of a pure substance. A pure substance, or more simply a **substance,** is matter that has a fixed composition. It can't be broken down into simpler parts by ordinary physical processes, such as boiling, grinding, or filtering. Only a chemical process can change a substance into one or more new substances. **Table 1** lists some examples of physical and chemical processes.

Reading Check *How can a substance be broken down?*

Table 1 Physical and Chemical Processes

Physical Processes	Chemical Processes
boiling	burning
changing pressure	reacting with other chemicals
cooling	reacting with light
sorting	

Compounds Water is another example of a substance. This is why water is always water even when you boil it or freeze it. Water, however, is not an element. It is an example of a compound. A compound is made of two or more elements that are chemically combined. Compounds also have fixed compositions. The ratio of the atoms in a compound is always the same. For example, when two hydrogen atoms combine with one oxygen atom, water is formed. All water—whether it's in the form of ice, liquid, or steam—has the same ratio of hydrogen atoms to oxygen atoms.

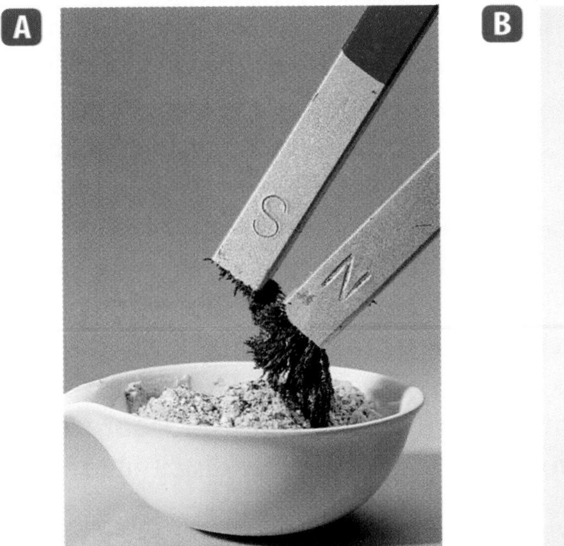

Figure 1
Mixtures can be separated by physical processes, such as by A attraction to a magnet or by B straining. *Why aren't the iron-sand mixture and the pulpy lemonade pure substances?*

Mixtures

Imagine drinking a glass of salt water. You would know right away that you weren't drinking pure water. Like salt water, many things are not pure substances. Salt water is a mixture of salt and water. Mixtures are combinations of substances that are not bonded together and can be separated by physical processes. For example, you can boil salt water to separate the salt from the water. If you had a mixture of iron filings and sand, you could separate the iron filings with a magnet. **Figure 1** shows some mixtures being separated.

Unlike compounds, mixtures do not always contain the same proportions of the substances that they are composed of. Lemonade is a mixture that can be strong tasting or weak tasting, depending on the amounts of water and lemon juice that are added. It also can be sweet or sour, depending on how much sugar is added. But whether it is strong, weak, sweet, or sour, it is still lemonade.

Heterogeneous Mixtures It is easy to tell that some things are mixtures just by looking at them. A watermelon is a mixture of fruit and seeds. The seeds are not evenly spaced through the whole melon—one bite you take might not have any seeds in it and another bite might have several seeds. This type of mixture, where the substances are not mixed evenly, is called a heterogeneous (he tuh ruh JEE nee us) mixture. The different areas of a heterogeneous mixture have different compositions. The substances in a heterogeneous mixture are usually easy to tell apart, like the seeds from the fruit of a watermelon. Other examples of heterogeneous mixtures include a bowl of cold cereal with milk and the mixture of pens, pencils, and books in your backpack.

Research Visit the Glencoe Science Web site at **science.glencoe.com** to find out how salt is removed from salt water to provide drinking water. Give a class presentation explaining at least two methods that you learned.

Homogeneous Mixtures Your shampoo contains many ingredients, but you can't see them when you look at the shampoo. It is the same color and texture throughout. Shampoo is an example of a homogeneous (hoh muh JEE nee us) mixture. A homogeneous mixture contains two or more substances that are evenly mixed on a molecular level but still are not bonded together. Another name for a homogeneous mixture is a **solution.** The sugar water in a hummingbird feeder, shown in **Figure 2,** is a solution—the sugar is evenly distributed in the water, and you can't see the sugar.

Figure 2
Molecules of sugar and water are evenly mixed in the solution in this hummingbird feeder.

Reading Check *What is another name for a homogeneous mixture?*

How Solutions Form

How do you make sugar water for a hummingbird feeder? You might add some sugar to a container of water and heat the mixture until the sugar disappeared. The sugar molecules would spread out until they were evenly spaced throughout the water, forming a solution. This is called dissolving. The substance that dissolves—or seems to disappear—is called the **solute.** The substance that dissolves the solute is called the **solvent.** In the hummingbird feeder solution, the solute is the sugar and the solvent is water. In the case of gas or solid solutions, the substance that is present in the greatest quantity is the solvent.

Precipitates Under certain conditions, a solid solute can come back out of its solution—or fall out—and form a solid. The solid is then called a **precipitate** (prih SIH puh tut), and this process is called precipitation. Sometimes, the precipitate will form if the solution is cooled or allowed to sit for a long time. Precipitates also can form because of a chemical reaction. In fact, precipitates probably have formed in your sink or shower because of chemical reactions. **Figure 3** illustrates the product of the reaction. Minerals that are dissolved in tap water react chemically with soap. The product of this reaction leaves the water as a precipitate called soap scum.

Figure 3
Minerals and soap combine to form soap scum, which falls out of the water solution and coats the tiles of a shower.

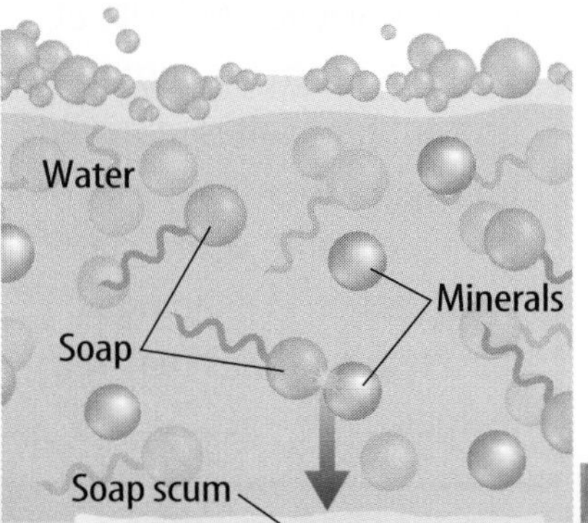

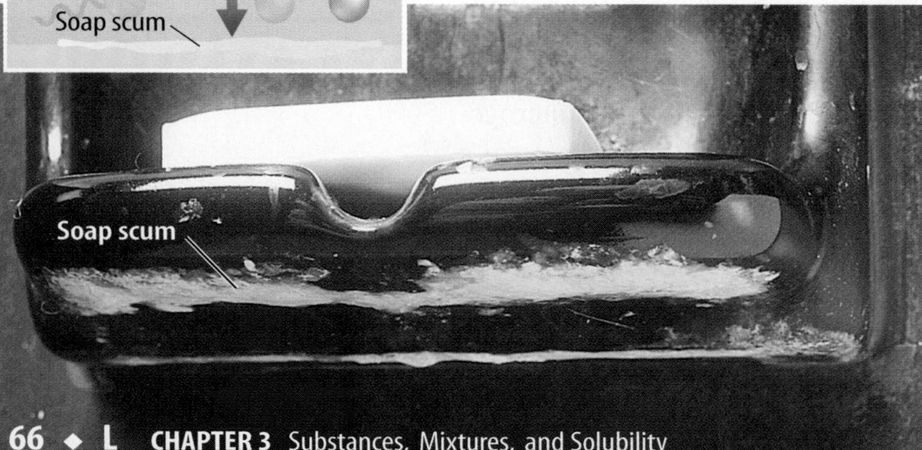

Ionic Bonds Some atoms do not share electrons when they join with other atoms to form compounds. Instead, these atoms lose or gain electrons. When they do, the number of protons and electrons within an atom are no longer equal, and the atom becomes positively or negatively charged. Atoms with a charge are called ions. Bonds that are formed by the transfer of electrons are called ionic bonds, and the compound that is formed is called an ionic compound. Table salt is an ionic compound that is made of sodium ions and chloride ions. Each sodium atom loses one electron to a chlorine atom and becomes a positively charged sodium ion. Each chlorine atom gains one electron from a sodium atom, becoming a negatively charged chloride ion.

Reading Check *How does an ionic compound differ from a molecular compound?*

Environmental Science INTEGRATION

Seawater is a solution that contains nearly every element found on Earth. Most elements are present in tiny quantities. Sodium and chloride ions are the most common ions in seawater. Several gases, including oxygen, nitrogen, and carbon dioxide also are dissolved in seawater.

How Water Dissolves Ionic Compounds Now think about the properties of water and the properties of ionic compounds as you visualize how an ionic compound dissolves in water. Because water molecules are polar, they attract positive and negative ions. The more positive part of a water molecule—where the hydrogen atoms are—is attracted to negatively charged ions. The more negative part of a water molecule—where the oxygen atom is—attracts positive ions. When an ionic compound is mixed with water, the different ions of the compound are pulled apart by the water molecules. **Figure 8** shows how sodium chloride dissolves in water.

Figure 8
Water dissolves table salt because its partial charges are attracted to the charged ions in the salt.

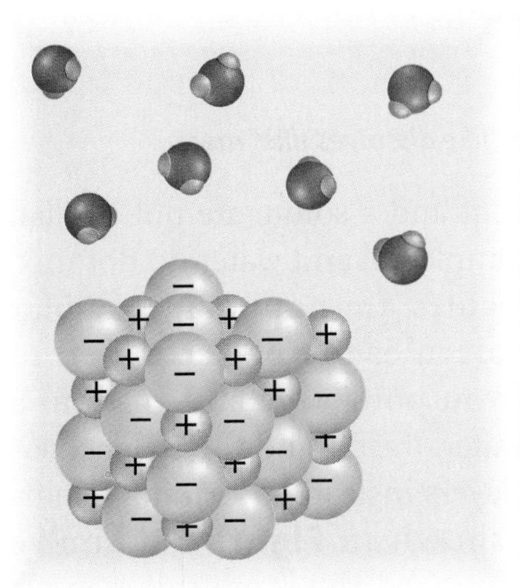

A The partially negative oxygen in the water molecule is attracted to a positive sodium ion.

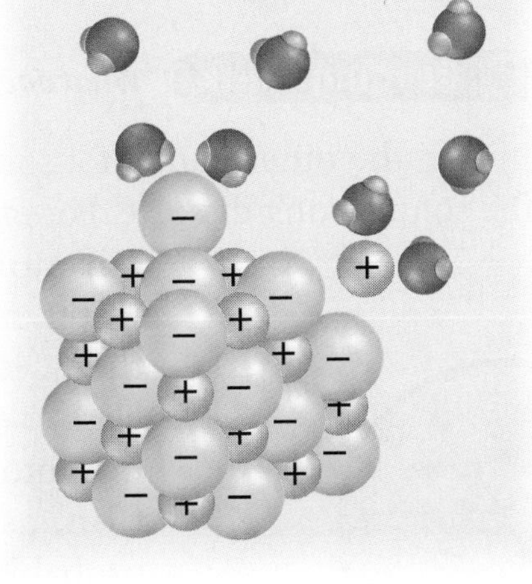

B The partially positive hydrogen atoms in another water molecule are attracted to a negative chloride ion.

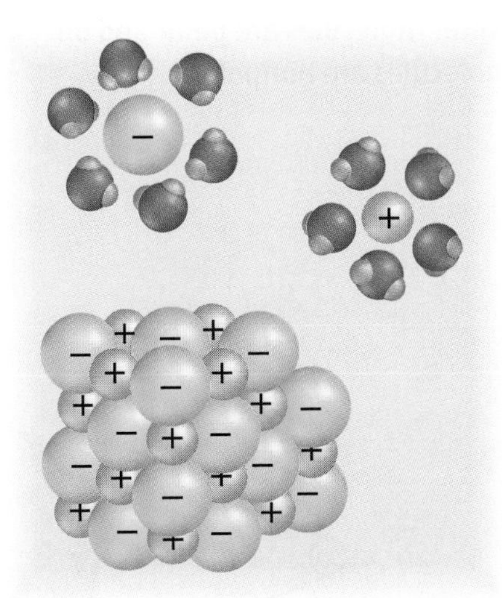

C The sodium and chloride ions are pulled apart from each other, and more water molecules are attracted to them.

How Water Dissolves Molecular Compounds

Can water also dissolve molecular compounds that are not made of ions? Water does dissolve molecular compounds, such as sugar, although it doesn't break each sugar molecule apart. Water simply moves between different molecules of sugar, separating them. Like water, a sugar molecule is polar. Polar water molecules are attracted to the positive and negative portions of the polar sugar molecules. When the sugar molecules are separated by the water and spread throughout it, as **Figure 9** shows, they have dissolved.

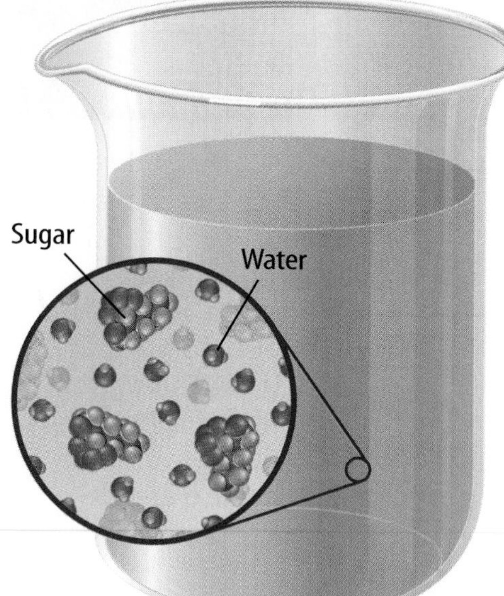

Figure 9
Sugar molecules that are dissolved in water spread out until they are spaced evenly in the water.

What will dissolve?

When you stir a spoonful of sugar into iced tea, all of the sugar dissolves but none of the metal in the spoon does. Why does sugar dissolve in water, but metal does not? A substance that dissolves in another is said to be soluble in that substance. You would say that the sugar is soluble in water but the metal of the spoon is insoluble in water, because it does not dissolve readily.

Like Dissolves Like When trying to predict which solvents can dissolve which solutes, chemists use the rule of "like dissolves like." This means that polar molecules dissolve polar molecules and nonpolar molecules dissolve nonpolar molecules. In the case of sugar and water, both are made up of polar molecules, so sugar is soluble in water. In the case of salt and water, the charged sodium and chloride ions are like the water molecule because it has a positive charge at one end and a negative charge at the other end.

Reading Check *What does "like dissolves like" mean?*

On the other hand, if a solvent and a solute are not similar, the solute won't dissolve. For example, oil and water do not mix. Oil molecules are nonpolar, so polar water molecules are not attracted to them. If you pour vegetable oil into a glass of water, the oil and the water separate into layers instead of forming a solution, as shown in **Figure 10.** You've probably noticed the same thing about the oil-and-water mixtures that make up some salad dressings. The oil stays on the top. Oils generally dissolve better in solvents that have nonpolar molecules.

Figure 10
Water and oil do not mix because water molecules are polar and oil molecules are nonpolar.

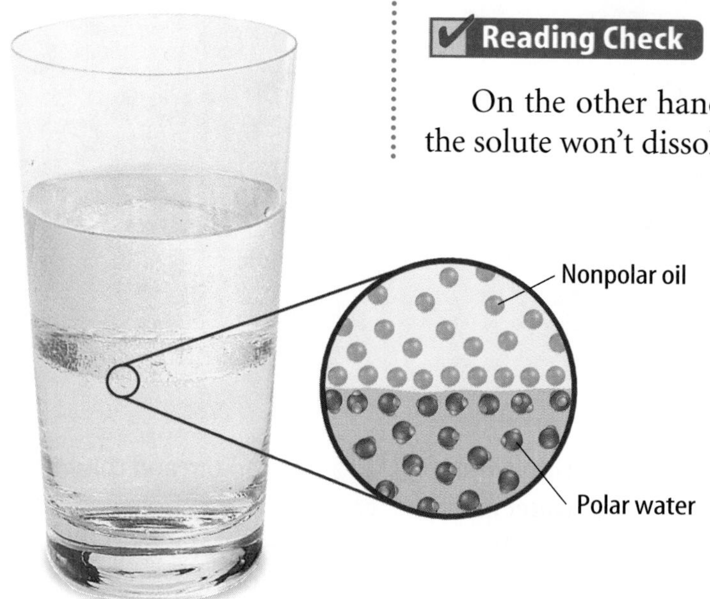

How much will dissolve?

Even though sugar is soluble in water, if you tried to dissolve 1 g of sugar into one small glass of water, not all of the sugar would dissolve. **Solubility** (sahl yuh BIH luh tee) is a measurement that describes how much solute dissolves in a given amount of solvent. The solubility of a material has been described as the amount of the material that can dissolve in 100 g of solvent at a given temperature. Some solutes are highly soluble, meaning that a large amount of solute can be dissolved in 100 g of solvent. For example, 63 g of potassium chromate can be dissolved in 100 g of water at 25°C. On the other hand, some solutes are not very soluble. For example, only 0.000 25 g of barium sulfate will dissolve in 100 g of water at 25°C. When a substance has an extremely low solubility, like barium sulfate does in water, it usually is considered insoluble.

Reading Check *What is an example of a substance that is considered to be insoluble?*

Rate of Dissolving Solubility does not tell you how fast a solute will dissolve—it tells you only how much of a solute will dissolve at a given temperature. Some solutes dissolve quickly, but others take a long time to dissolve. A solute dissolves faster when the solution is stirred or shaken or when the temperature of the solution is increased. These methods increase the rate at which the surfaces of the solute come into contact with the solvent. Increasing the area of contact between the solute and the solvent can also increase the rate of dissolving. This can be done by breaking up the solute into smaller pieces, which increases the surface area of the solute that is exposed to the solvent.

Solubility in Liquid-Solid Solutions Did you notice that the temperature was included in the explanation about the amount of solute that dissolves in a quantity of solvent? The solubility of many solutes changes if you change the temperature of the solvent. For example, if you heat water, not only does the sugar dissolve at a faster rate, but more sugar can dissolve in it. However, some solutes, like salt and calcium carbonate, do not become more soluble when the temperature of water increases. The graph in **Figure 11** shows how the temperature of the solvent affects the solubility of some solutes.

Figure 11
The solubility of some solutes changes as the temperature of the solvent increases. *According to the graph, is it likely that warm ocean water contains any more salt than cold ocean water does?*

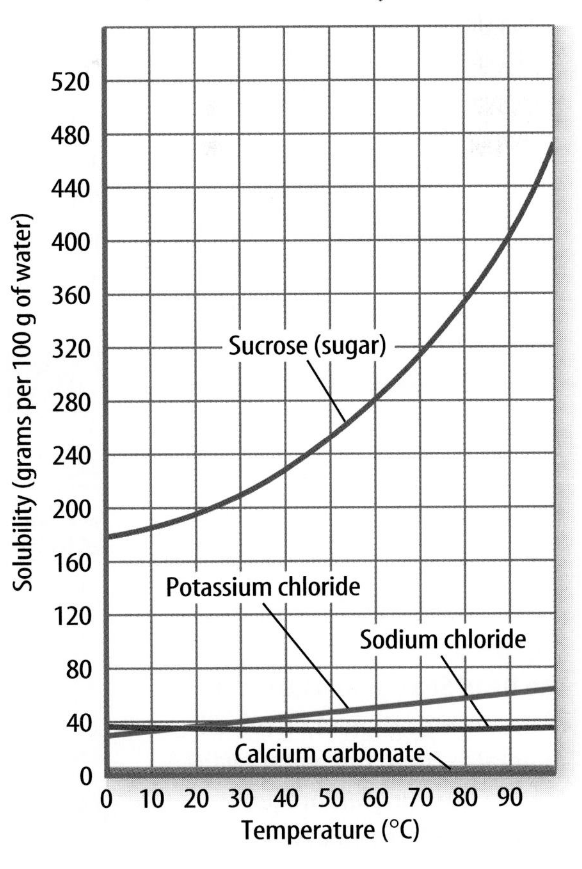

Observing Gas Solubility

Procedure

1. Obtain a bottle of a thoroughly chilled, **carbonated beverage.**
2. Carefully remove the cap from the bottle with as little agitation as possible.
3. Quickly cover the opening with an uninflated **balloon.** Use **tape** to secure and tightly seal the balloon to the top of the bottle.
4. **WARNING:** *Be careful not to point the bottle at anyone.* Gently agitate the bottle from side to side for 2 min and observe the balloon.
5. Set the bottle of soft drink in a **container** of hot tap **water** for 10 min and observe the balloon.

Analysis

1. Compare and contrast the amounts of carbon dioxide gas released from the cold and the warm soft drinks.
2. Why does the soft drink contain a different amount of carbon dioxide when it is warmed than it does when it is chilled?

Solubility in Liquid-Gas Solutions Unlike liquid-solid solutions, an increase in temperature decreases the solubility of a gas in a liquid-gas solution. You might notice this if you have ever opened a warm carbonated beverage and it bubbled up out of control while a chilled one barely fizzed. Carbon dioxide is less soluble in a warm solution. What keeps the carbon dioxide from bubbling out when it is sitting at room temperature on a supermarket shelf? When a bottle is filled, extra carbon dioxide gas is squeezed into the space above the liquid increasing the pressure in the bottle. This increased pressure increases the solubility of gas and forces most of it into solution. When you open the cap, the pressure is released and the solubility of the carbon dioxide decreases.

Saturated Solutions If you add calcium carbonate to 100 g of water at 25°C, only 0.001 4 g of it will dissolve. Additional calcium carbonate will not dissolve. Such a solution—one that contains all of the solute that it can hold under the given conditions—is called a **saturated** solution. **Figure 12** shows a saturated solution. If a solution is a solid-liquid solution, the extra solute that is added will settle to the bottom of the container. Of course, it's possible to make solutions that have less solute than they would need to become saturated. Such solutions are unsaturated. An example of an unsaturated solution is one containing 50 g of sugar in 100 g of water at 25°C. That's much less than the 204 g of sugar the solution would need to be saturated.

A hot solvent usually can hold more solute than a cool solvent can. When a saturated solution cools, some of the solute usually falls out of the solution. But if a saturated solution is cooled slowly, sometimes the excess solute remains dissolved for a period of time. Such a solution is said to be supersaturated, because it contains more than the normal amount of solute.

Figure 12
The Dead Sea has an extremely high concentration of dissolved minerals. When the water evaporates, the minerals are left behind and form pillars.

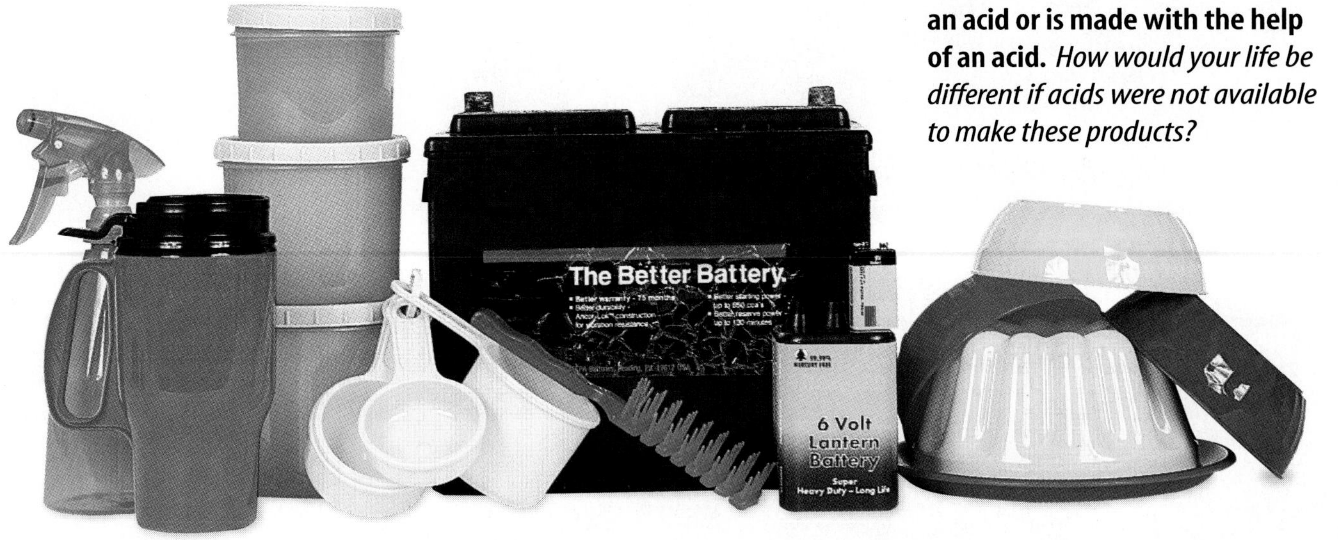

Figure 15
Each of these products contains an acid or is made with the help of an acid. *How would your life be different if acids were not available to make these products?*

Uses of Acids You're probably familiar with many acids. Vinegar, which is used in salad dressing, contains acetic acid. Lemons, limes, and oranges have a sour taste because they contain citric acid. Your body needs ascorbic acid, which is vitamin C. Ants that sting inject formic acid into their victims.

Figure 15 shows other products that are made with acids. Sulfuric acid is used in the production of fertilizers, steel, paints, and plastics. Acids often are used in batteries because their solutions conduct electricity. For this reason, it sometimes is referred to as battery acid. Hydrochloric acid, which is known commercially as muriatic acid, is used in a process called pickling. Pickling is a process that removes impurities from the surfaces of metals. Hydrochloric acid also can be used to clean mortar from brick walls. Nitric acid is used in the production of fertilizers, dyes, and plastics.

Acid in the Environment Carbonic acid plays a key role in forming cave formations and in the formation of stalactites and stalagmites. Carbonic acid is formed when carbon dioxide in soil is dissolved in water. When this acidic solution comes in contact with calcium carbonate—or limestone rock—it can dissolve it, eventually carving out a cave in the rock. A similar process occurs when acid rain falls on statues and eats away at the stone, as shown in **Figure 16.** When this acidic solution drips from the ceiling of the cave, water evaporates and carbon dioxide becomes less soluble, forcing it out of solution. The solution becomes less acidic and the limestone becomes less soluble, causing it to precipitate out of solution. These precipitates form stalactites and stalagmites.

TRY AT HOME
Mini LAB

Observing a Nail in a Carbonated Drink

Procedure

1. Observe the initial appearance of an **iron nail.**
2. Pour enough **carbonated soft drink** into a **cup or beaker,** to cover the nail.
3. Drop the nail into the soft drink and observe what happens.
4. Leave the nail in the soft drink overnight and observe it again the next day.

Analysis

1. Describe what happened when you first dropped the nail into the soft drink and the appearance of the nail the following day.
2. Based upon the fact that the soft drink was carbonated, explain why you think the drink reacted with the nail as you observed.

VISUALIZING ACID PRECIPITATION

Figure 16

When fossil fuels such as coal and oil are burned, a variety of chemical compounds are produced and released into the air. In the atmosphere, some of these compounds form acids that mix with water vapor and fall back to Earth as acid precipitation—rain, sleet, snow, or fog. The effects of acid precipitation on the environment can be devastating. Winds carry these acids hundreds of miles from their source, damaging forests, corroding statues, and endangering human health.

B Sulfur dioxide and nitrogen oxides react with water vapor in the air to form highly acidic solutions of nitric acid (HNO_2) and sulfuric acid (H_2SO_4). These solutions eventually return to Earth as acid precipitation.

C Some acid rain in the United States has a pH of 2.0 or lower—about the acidity of stomach acid.

A Power plants and cars burn fossil fuels to generate energy for human use. In the process, sulfur dioxide (SO_2) and nitrogen oxides are released into the atmosphere.

Procedure

1. **Design** a data table to record the names of the solutions to be tested, the colors caused by the added cabbage juice indicator, and the relative pH values of the solutions.
2. Mark each test tube with the identity of the acid or base solution it will contain.
3. Half-fill each test tube with the solution to be tested.
   WARNING: *If you spill any liquids on your skin, rinse the area immediately with water. Alert your teacher if any liquid is spilled in the work area.*
4. Add ten drops of the cabbage juice indicator to each of the solutions to be tested. Gently agitate or wiggle each test tube to mix the cabbage juice with the solution.
5. **Observe** and record the color of each solution in your data table.

Determining pH Values

Cabbage Juice Color	Relative Strength of Acid or Base
	strong acid
	medium acid
	weak acid
	neutral
	weak base
	medium base
	strong base

Conclude and Apply

1. **Compare** your observations with the table above. Record in your data table the relative pH of each solution you tested.
2. **Classify** which solutions were acidic and which were basic.
3. Which solution was the weakest acid? The strongest base? The closest to neutral pH?
4. **Predict** what ion might be involved in the cleaning process based upon your pH values for the ammonia, soap, and borax soap solutions.
5. Form a hypothesis that explains why the borax soap solution had a much higher pH than an ammonia solution of approximately the same concentration.

Use your data to **create** labels for the solutions you tested. Include the relative pH of each solution and any other safety information you think is important on each label. **For more help, refer to the** Science Skill Handbook.

Salty Solutions

Did you know...

...Seawater is certainly a salty solution. It contains almost 60 chemical elements, but the most common ingredient is sodium chloride (table salt). Seawater also contains trace amounts of rare metals, including silver, copper, uranium, and gold.

...Tears and saliva have a lot in common. Both are salty solutions that protect you from harmful bacteria, keep tissues moist, and help spread nutrients. Bland-tasting saliva, however, is 99.5 percent water. The remaining 0.5 percent is a combination of many ions, including sodium, and several proteins.

...Your body is a little like an ocean. It contains a solution similar to seawater. The body of a 60-kg person contains about 41 L of salt water.

...The largest salt lake in the United States is the Great Salt Lake. It covers more than 4,000 km^2 in Utah and is up to 13.4 m deep. The Great Salt Lake and the Salt Lake Desert were once part of the enormous, prehistoric Lake Bonneville, which was 305 m deep at some points.

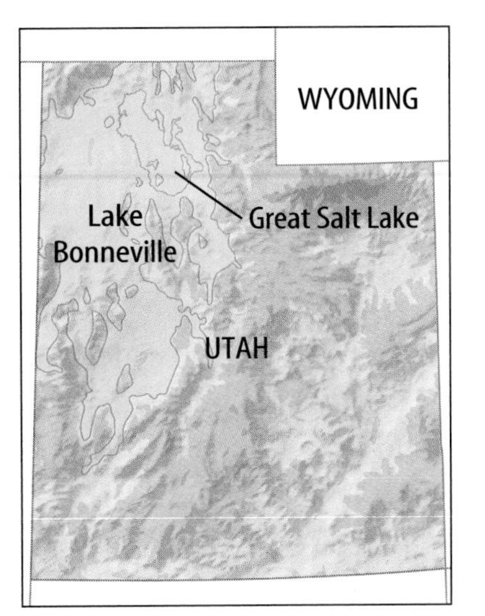

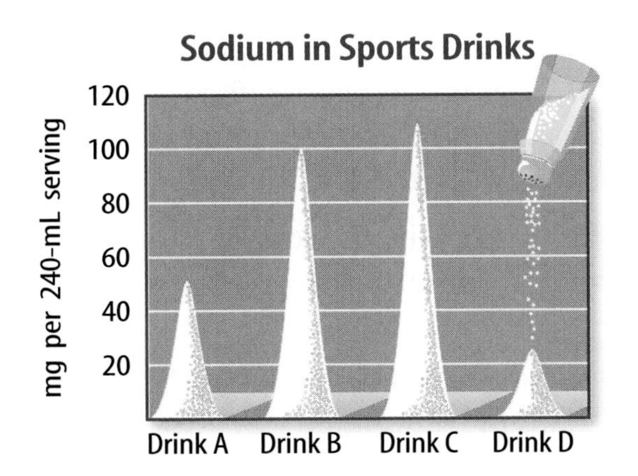

...Salt can reduce pain. Gargled salt water is a disinfectant; it fights the bacteria that cause sore throats.

Do the Math

1. Some athletes use pickle juice as a sports drink—0.06 L of it contain 700 mg of sodium. About how many milliliters of sports drink C would it take to equal this amount of sodium?
2. At its largest, Lake Bonneville covered about 32,000 km^2. What percentage of that area does the Great Salt Lake now cover?
3. Seawater has a density of about 1.03 kg/L. If this is similar to the density of the salty solution in the human body, about what percentage of a 60-kg person's mass is this salty solution?

Go Further

Do research to learn about other elements in seawater. Create a graph that shows the amounts of the ten most common elements in 1 L of seawater.

Chapter 3 Study Guide

Reviewing Main Ideas

Section 1 What is a solution?

1. Elements and compounds are pure substances, because their compositions are fixed. Mixtures are not pure substances.
2. Heterogeneous mixtures are not mixed evenly. Homogeneous mixtures, also called solutions, are mixed evenly on a molecular level. In a solution, the components cannot be distinguished easily. *What parts of chicken noodle soup are heterogeneous? What parts are homogeneous?*

3. When solutes dissolve in solvents, solutions are formed.
4. Solutes and solvents can be gases, liquids, or solids, combined in many different ways.

Section 2 Solubility

1. Because water molecules are polar, they can dissolve many different solutes, including ionic compounds and polar molecular compounds. Like dissolves like.
2. The maximum amount of solute that can be dissolved in a given amount of solvent at a particular temperature is called the solubility of the solute. Temperature and pressure can affect solubility.
3. Solutions can be unsaturated, saturated, or supersaturated, depending on how much solute is dissolved compared to the solubility of the solute in the solvent.
4. The concentration of a solution is the amount of solute in a particular volume of solvent. *Which fruit drink is more concentrated?*

Section 3 Acidic and Basic Solutions

1. Acids release H^+ ions and produce hydronium ions when they are dissolved in water. Bases accept H^+ ions and produce hydroxide ions when dissolved in water.
2. Acidic and basic solutions have distinctive properties, such as being corrosive and being able to conduct electricity.
3. pH expresses the concentrations of hydronium ions and hydroxide ions in aqueous solutions. Solutions with pH values below 7 are acidic. Solutions with a pH of 7 are neutral. Solutions with pH values above 7 are basic. *How does this litmus paper test pH?*

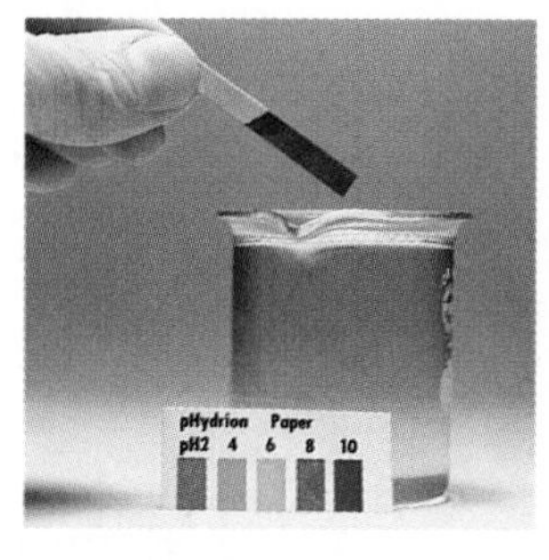

4. In a neutralization reaction, an acid reacts with a base to form water and a salt.

After You Read

FOLDABLES Reading & Study Skills

Using what you have learned, circle the solutions that are acids on your Foldable and underline the solutions that are bases.

Visualizing Main Ideas

Complete the concept map on the classification of matter.

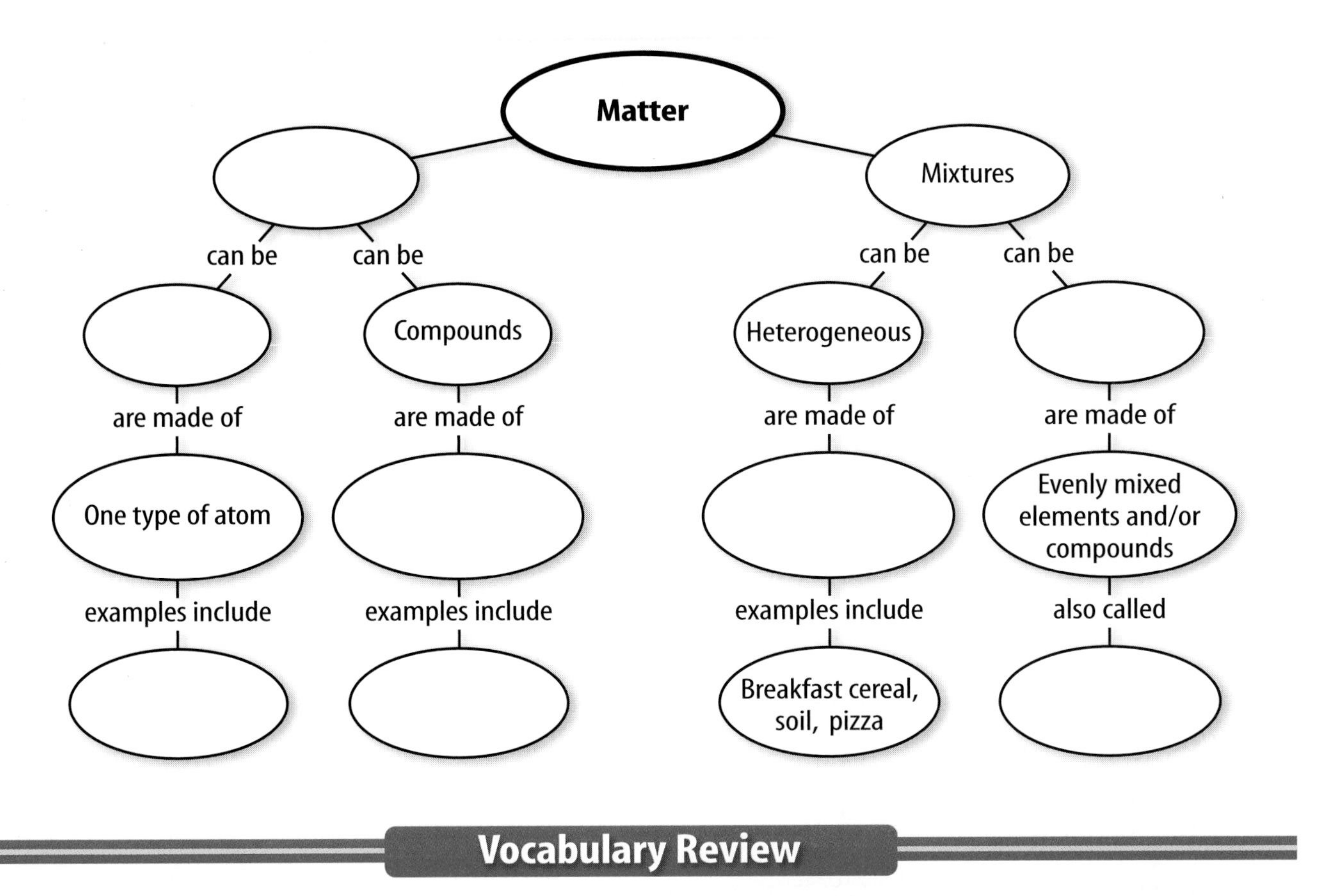

Vocabulary Review

Vocabulary Words

a. acid
b. aqueous
c. base
d. concentration
e. indicator
f. neutralization
g. pH
h. precipitate
i. saturated
j. solubility
k. solute
l. solution
m. solvent
n. substance

Study Tip

Study with a friend. Quiz each other from textbook and class material. Take turns asking questions and giving answers.

Using Vocabulary

Each of the following sentences is false. Make the sentence true by replacing the underlined word with the correct vocabulary word.

1. A base has a <u>precipitate</u> value above 7.
2. A measure of how much solute is in a solution is its <u>solubility</u>.
3. The amount of a solute that can dissolve in 100 g of solvent is its <u>pH</u>.
4. The <u>solvent</u> is the substance that is dissolved to form a solution.
5. The reaction between an acidic and a basic solution is called <u>concentration</u>.
6. A <u>solution</u> has a fixed composition.

Chapter 3 Assessment

Checking Concepts

Choose the word or phrase that best answers the question.

1. Which of the following is a solution?
A) water
B) an oatmeal-raisin cookie
C) copper
D) vinegar

2. What type of compounds will not dissolve in water?
A) polar **C)** nonpolar
B) ionic **D)** charged

3. When chlorine compounds are dissolved in pool water, what is the water?
A) the alloy **C)** the solution
B) the solvent **D)** the solute

4. A solid might become less soluble in a liquid when you decrease what?
A) stirring **C)** temperature
B) pressure **D)** container size

5. Which acid is used in the industrial process known as pickling?
A) hydrochloric **C)** sulfuric
B) carbonic **D)** nitric

6. When few particles of solute are in a large amount of solvent, what kind of solution is formed?
A) soluble **C)** concentrated
B) dilute **D)** insoluble

7. Given equal concentrations, which of the following will produce the most hydronium ions in an aqueous solution?
A) a strong base **C)** a strong acid
B) a weak base **D)** a weak acid

8. Bile, a body fluid involved in digestion, is acidic—what is its pH?
A) 11 **C)** less than 7
B) 7 **D)** greater than 7

9. What type of molecule is water?
A) polar **C)** nonpolar
B) ionic **D)** precipitate

10. When you swallow an antacid, what happens to your stomach acid?
A) It is more acidic. **C)** It is diluted.
B) It is concentrated. **D)** It is neutralized.

Thinking Critically

11. Why do deposits form in the steam vents of irons in some parts of the country?

12. Is it possible to have a dilute solution of a strong acid? Explain.

13. In which type of container is orange juice more likely to keep its freshly squeezed taste—a metal can or a paper carton? Explain.

14. A chemistry professor always adds sugar to his coffee before adding milk. Can you suggest why he might want to do this?

15. Water molecules can break apart to form H^+ ions and OH^- ions. Water is known as an amphoteric substance, which is something that can act as an acid or a base. Explain how this can be so.

Developing Skills

16. Recognizing Cause and Effect When you slice an apple, the cut surfaces soon turn brown. Brushing fruit with lemon juice slows this darkening process. Explain why this might occur in terms of the effect of pH on this reaction.

17. **Making and Using Graphs** Using the solubility graph below, estimate the solubilities of potassium chloride and sodium chloride in grams per 100 g of water at 80°C.

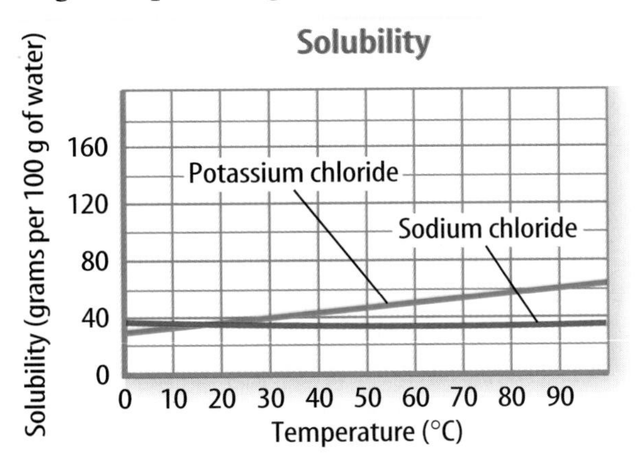

18. **Comparing and Contrasting** Give examples of heterogeneous and homogeneous mixtures from your daily life and compare and contrast them.

19. **Forming Hypotheses** A warm carbonated beverage seems to fizz more than a cold one when it is opened. Explain this based on the solubility of carbon dioxide in water.

Performance Assessment

20. **Illustrate** Make a poster of a pH scale using pictures of solutions cut from magazines or drawn by hand. Use reference materials to find the pH of the items you have chosen, and place them correctly on your pH scale.

21. **Poem** Write a poem that explains the difference between a substance and a mixture.

Technology

Go to the Glencoe Science Web site at **science.glencoe.com** or use the **Glencoe Science CD-ROM** for additional chapter assessment.

Test Practice

Mixtures are made when different substances are mixed but can be physically separated again. Some heterogeneous and homogeneous mixtures have been divided into the two boxes below.

Study the diagram and answer the following questions.

1. According to the pictures, a heterogeneous mixture is a mixture that _____ .
   A) contains only one liquid
   B) is made up of food
   C) shows the different substances easily
   D) is mixed evenly at the molecular level

2. Which of these belongs with the homogeneous mixture group above?
   F) maple syrup
   G) rice and beans
   H) a blueberry muffin
   J) mixed nuts

CHAPTER

4

Carbon Chemistry

When you are camping, you spend your days outdoors and your nights sleeping under the stars—activities that are different from your daily life. But one thing stays the same—carbon compounds surround you. Your food, clothing, shoes, cooler, tent, body, and the other living things around you contain carbon compounds. All living things and many manufactured items are made from carbon compounds. In this chapter, you will learn that because of carbon's atomic structure, it can form many different compounds.

What do you think?

Science Journal Look at the picture below with a classmate. Discuss what you think this might be. Here's a hint: *It's crunchy.* Write your answer or best guess in your Science Journal.

Many of the compounds that compose your body and other living things are carbon compounds. This activity demonstrates some of the atomic combinations possible with one carbon and four other atoms.

Model carbon's bonding

WARNING: *Do not eat any foods from this activity. Wash your hands before and after this activity.*

1. Insert four toothpicks into a small clay or plastic foam ball so they are evenly spaced around the sphere. The ball represents a carbon atom. The toothpicks represent chemical bonds.
2. Make models of as many molecules as possible by adding raisins, grapes, and gumdrops to the ends of the toothpicks. Use raisins to represent hydrogen atoms, grapes to represent chlorine atoms, and gumdrops to represent fluorine atoms.

Observe

Draw each model and write the formula for it in your Science Journal. What can you infer about the number of compounds a carbon atom can form?

Before You Read

Making a Vocabulary Study Fold **Knowing the definition of vocabulary words is a good way to ensure that you understand the content of the chapter.**

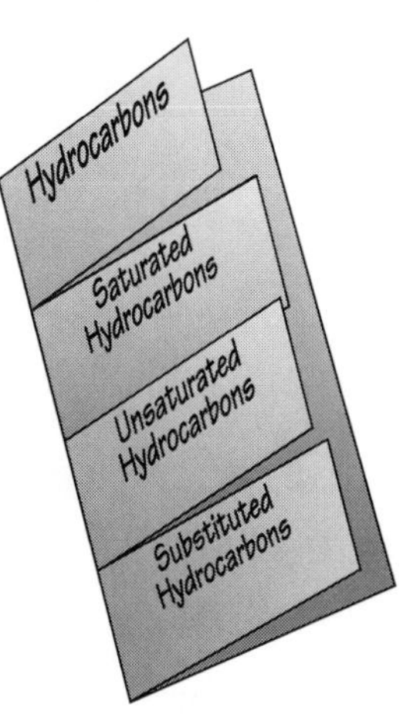

1. Place a sheet of paper in front of you so the short side is at the top. Fold the paper in half from the left side to the right side.
2. Now fold the paper in half from top to bottom. Then fold it in half again top to bottom. Unfold the last two folds.
3. Through the top thickness of paper, cut along each of the fold lines to form four tabs. Label the tabs *Hydrocarbons, Saturated Hydrocarbons, Unsaturated Hydrocarbons,* and *Substituted Hydrocarbons.*
4. As you read the chapter, find the definitions and write them on the front of the tabs.

SECTION

Simple Organic Compounds

As You Read

***What* You'll Learn**

- **Explain** why carbon is able to form many compounds.
- **Describe** how saturated and unsaturated hydrocarbons differ.
- **Identify** isomers of organic compounds.

Vocabulary

organic compound
hydrocarbon
saturated hydrocarbon
unsaturated hydrocarbon
isomer

***Why* It's Important**

Plants, animals, and many of the things that are part of your life are made of organic compounds.

Organic Compounds

Earth's crust contains less than one percent carbon, yet all living things on Earth are made of carbon-containing compounds. Carbon's ability to bond easily and form compounds is the basis of life on Earth. A carbon atom has four electrons in its outer energy level, so it can form covalent bonds with as many as four other atoms. When carbon atoms form covalent bonds, they obtain the stability of a noble gas with eight electrons in their outer energy level. One of carbon's most frequent partners in forming covalent bonds is hydrogen.

Substances can be classified into two groups—those derived from living things and those derived from nonliving things, as shown in **Figure 1.** Most of the substances associated with living things contain carbon and hydrogen. These substances were called organic compounds, which means "derived from a living organism." However, in 1828, scientists discovered that living organisms are not necessary to form organic compounds. Despite this, scientists still use the term **organic compounds** for most compounds that contain carbon.

Reading Check *What is the origin of the term* organic compound?

Figure 1
Most substances can be classified as living or nonliving things.

A **Living things and products made from living things such as this wicker chair contain carbon.**

B **Most of the things in this photo are nonliving and are composed of elements other than carbon.**

Hydrocarbons

Many compounds are made of carbon and hydrogen alone. A compound that contains only carbon and hydrogen atoms is called a **hydrocarbon.** The simplest hydrocarbon is methane, the primary component of natural gas. If you have a gas stove or gas furnace in your home, methane usually is the fuel that is burned in these appliances. Methane consists of a single carbon atom covalently bonded to four hydrogen atoms. The formula for methane is CH_4. **Figure 2** shows a model of the methane molecule and its structural formula. In a structural formula, the line between one atom and another atom represents a pair of electrons shared between the two atoms. This pair forms a single bond. Methane contains four single bonds.

Now, visualize the removal of one of the hydrogen atoms from a methane molecule, as in **Figure 3A.** A fragment of the molecule called a methyl group, $-CH_3$, would remain. The methyl group then can form a single bond with another methyl group. If two methyl groups bond with each other, the result is the two-carbon hydrocarbon ethane, C_2H_6, which is shown with its structural formula in **Figure 3B.**

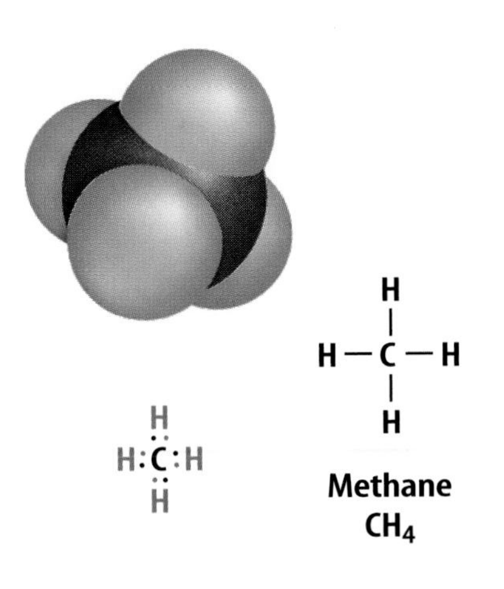

Figure 2
Methane is the simplest hydrocarbon molecule. *Why is this true?*

Figure 3
Here's a way to visualize how larger hydrocarbons are built up.

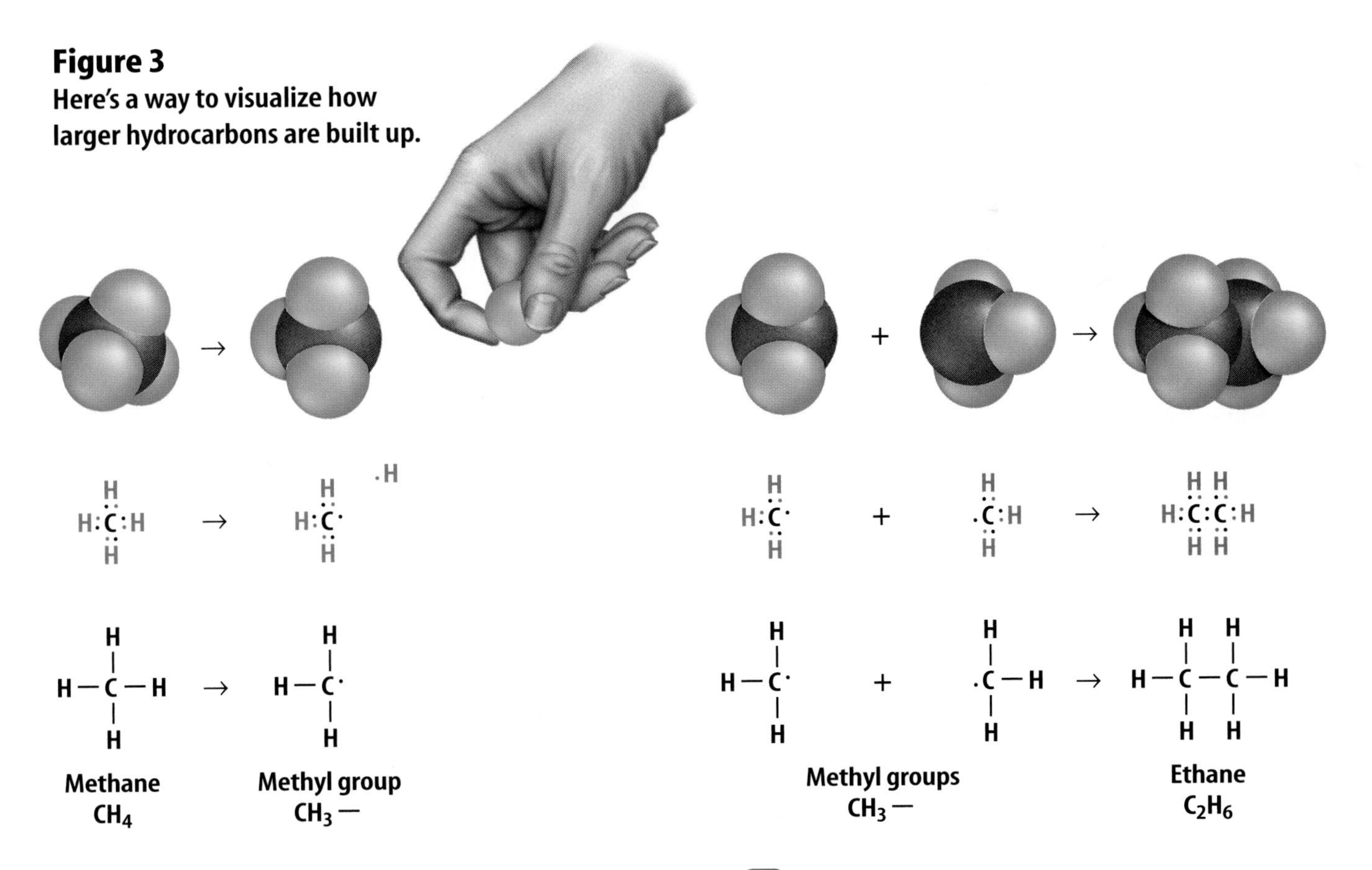

A **A hydrogen is removed from a methane molecule, forming a methyl group. A methyl group is a carbon atom bonded to three hydrogen atoms.**

B **Each carbon atom in ethane has four bonds after the two methyl groups join.**

Figure 4
Propane and butane are two useful fuels.
Why are they called "saturated"?

```
    H   H   H
    |   |   |
H — C — C — C — H
    |   |   |
    H   H   H
```

Propane
C_3H_8

A When propane burns, it releases energy as the chemical bonds are broken. Propane often is used to fuel camp stoves and outdoor grills.

```
    H   H   H   H
    |   |   |   |
H — C — C — C — C — H
    |   |   |   |
    H   H   H   H
```

Butane
C_4H_{10}

B Butane also releases energy when it burns. Butane is the fuel that is used in disposable lighters.

Saturated Hydrocarbons Methane and ethane are members of a family of molecules in which carbon and hydrogen atoms are joined by single covalent bonds. When all the bonds in a hydrocarbon are single bonds, the molecule is called a **saturated hydrocarbon.** They are called *saturated* because no additional hydrogen atoms can be added to the molecule. The carbon atoms are saturated with hydrogen atoms. The formation of larger hydrocarbons occurs in a way similar to the formation of ethane. A hydrogen atom is removed from ethane and replaced by a $-CH_3$ group. Propane, with three carbon atoms, is the third member of the series of saturated hydrocarbons. Butane has four carbon atoms. Both of these hydrocarbons are shown in **Figure 4.** The names and the chemical formulas of a few of the smaller saturated hydrocarbons are listed in **Table 1.** Saturated hydrocarbons are named with an *-ane* ending.

Reading Check *What is a saturated hydrocarbon?*

These short hydrocarbon chains have low boiling points, so they evaporate and burn easily. That makes methane a good fuel for your stove or furnace. Propane is used in gas grills, lanterns, and to heat air in hot-air balloons. Butane often is used as a fuel for camp stoves and lighters. Longer hydrocarbons are found in oils and waxes. Carbon can form long chains that contain hundreds or even thousands of carbon atoms. These extremely long chains make up many of the plastics that you use.

Earth Science INTEGRATION

Petroleum is a mixture of hydrocarbons that formed from plants and animals that lived in seas and lakes hundreds of millions of years ago. With the right temperature and pressure, this plant and animal matter, buried deep under Earth's surface, decomposed to form petroleum. Why is petroleum a nonrenewable resource?

Unsaturated Hydrocarbons Carbon also forms hydrocarbons with double and triple bonds. In a double bond, two pairs of electrons are shared between two atoms, and in a triple bond, three pairs of electrons are shared. Hydrocarbons with double or triple bonds are called **unsaturated hydrocarbons.** This is because the carbon atoms are not saturated with hydrogen atoms.

Ethene, the simplest unsaturated hydrocarbon, has two carbon atoms joined by a double bond. Propene is an unsaturated hydrocarbon with three carbons. Some unsaturated hydrocarbons have more than one double bond. Butadiene (byew tuh DI een) has four carbon atoms and two double bonds. The structures of ethene, propene, and butadiene are shown in **Figure 5.**

Unsaturated compounds with at least one double bond are named with an *-ene* ending. Notice that the names of the compounds below have an *-ene* ending.

Reading Check ***What type of bonds are found in unsaturated hydrocarbons?***

Table 1 The Structures of Hydrocarbons

Name	Structural Formula	Chemical Formula
Methane	H | H—C—H | H	CH_4
Ethane	H H | | H—C—C—H | | H H	C_2H_6
Propane	H H H | | | H—C—C—C—H | | | H H H	C_3H_8
Butane	H H H H | | | | H—C—C—C—C—H | | | | H H H H	C_4H_{10}

Figure 5
You'll find unsaturated hydrocarbons in many of the products you use every day.

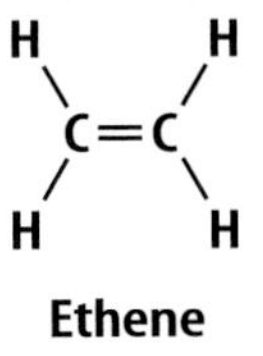

Ethene

A **Ethene helps ripen fruits and vegetables. It's also used to make milk and soft-drink bottles.**

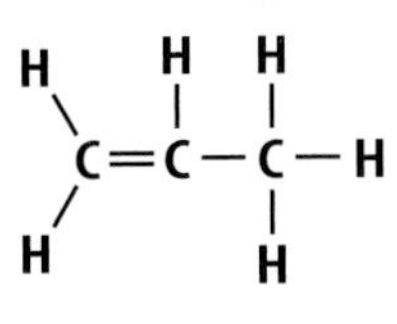

Propene

B **This detergent bottle contains the tough plastic polypropylene, which is made from propene.**

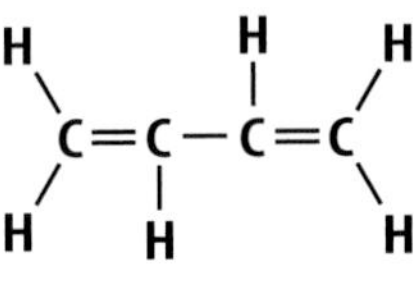

Butadiene

C **Butadiene made it possible to replace natural rubber with synthetic rubber.**

Figure 6
In the welder's torch, ethyne, or acetylene, is combined with oxygen to form a mixture that burns, releasing intense light and heat. The two carbon atoms in ethyne are joined by a triple bond.

H—C≡C—H
Ethyne or Acetylene
C_2H_2

Modeling Isomers

Procedure
WARNING: *Do not eat any foods in this activity.*
1. Construct a model of pentane, C_5H_{12}. Use **toothpicks** for covalent bonds and small balls of different **colored clay or gumdrops** for carbon and hydrogen atoms.
2. Using the same materials, build a molecule with a different arrangement of the atoms. Are there any other possibilities?
3. Make a model of hexane, C_6H_{14}.
4. Arrange the atoms of hexane in different ways.

Analysis
1. How many isomers of pentane did you build? How many isomers of hexane?
2. Do you think there are more isomers of heptane, C_7H_{16}, than hexane? Explain.

Triple Bonds Unsaturated hydrocarbons also can have triple bonds, as in the structure of ethyne (EH thine) shown in **Figure 6.** Ethyne, commonly called acetylene (uh SE tul een), is a gas used for welding because it produces high heat as it burns. Welding torches mix acetylene and oxygen before burning.

Hydrocarbon Isomers Suppose you had ten blocks that could be snapped together in different combinations. Each combination of the same ten blocks is different. The atoms in an organic molecule also can have different arrangements but still have the same molecular formula. Compounds that have the same molecular formula but different structures are called **isomers** (I suh murz). Two isomers, butane and isobutane, are shown in **Figure 7.** They have different chemical and physical properties because of their different structures. As the size of a hydrocarbon molecule increases, the number of possible isomers also increases.

By now, you might be confused about how organic compounds are named. **Figure 8** explains the system that is used to name simple organic compounds.

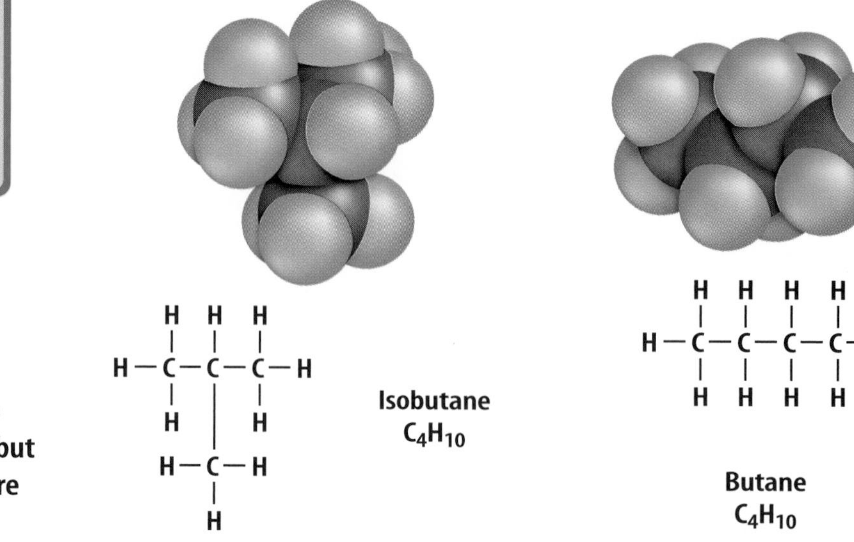

Figure 7
Butane and isobutane both have four carbons and ten hydrogens but their structures and properties are different.

Figure 8

Millions of organic compounds have been discovered and created, and thousands of new ones are synthesized in laboratories every year. To keep track of all these carbon-containing molecules, the International Union of Pure and Applied Chemistry, or IUPAC, devised a special naming system (a nomenclature) for organic compounds. As shown here, different parts of an organic compound's name—its root, suffix, or prefix—give information about its size and structure.

Carbon atoms	Name	Molecular formula
1	Methane	CH_4
2	Ethane	CH_3CH_3
3	Propane	$CH_3CH_2CH_3$
4	Butane	$CH_3CH_2CH_2CH_3$
5	Pentane	$CH_3CH_2CH_2CH_2CH_3$
6	Hexane	$CH_3CH_2CH_2CH_2CH_2CH_3$
7	Heptane	$CH_3CH_2CH_2CH_2CH_2CH_2CH_3$
8	Octane	$CH_3CH_2CH_2CH_2CH_2CH_2CH_2CH_3$
9	Nonane	$CH_3CH_2CH_2CH_2CH_2CH_2CH_2CH_2CH_3$
10	Decane	$CH_3CH_2CH_2CH_2CH_2CH_2CH_2CH_2CH_2CH_3$

ROOT WORDS The key to every name given to a compound in organic chemistry is its root word. This word tells how many carbon atoms are found in the longest continuous carbon chain in the compound. Except for compounds with one to four carbon atoms, the root word is based on Greek numbers.

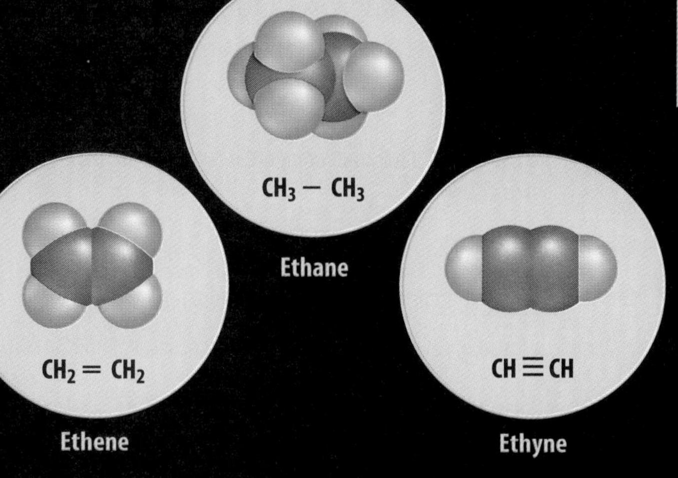

SUFFIXES The suffix of the name for an organic compound indicates the kind of covalent bonds joining the compound's carbon atoms. If the atoms are joined by single covalent bonds, the compound's name will end in *-ane*. If there is a double covalent bond in the carbon chain, the compound's name ends in *-ene*. Similarly, if there is a triple bond in the chain, the compound's name will end in *-yne*.

PREFIXES The prefix of the name for an organic compound describes how the carbon atoms in the compound are arranged. Organic molecules that have names with the prefix *cyclo-* contain a ring of carbon atoms. For example, cyclopentane contains five carbon atoms all joined by single bonds in a ring.

Cyclopentane

Figure 9
Visualize a hydrogen atom removed from a carbon atom on both ends of a hexane chain. The two end carbons form a bond with each other. *How does the chemical formula change?*

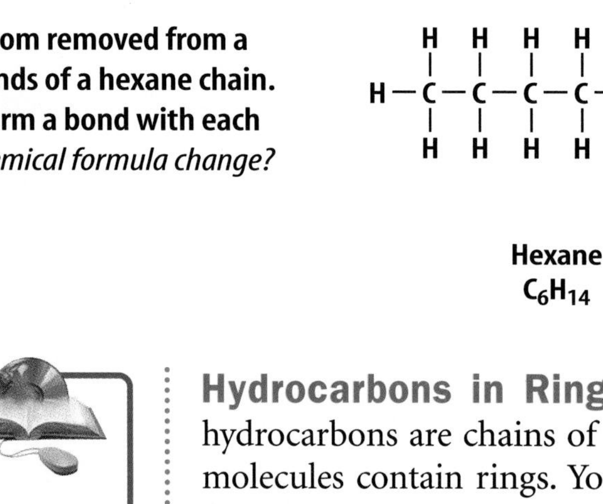

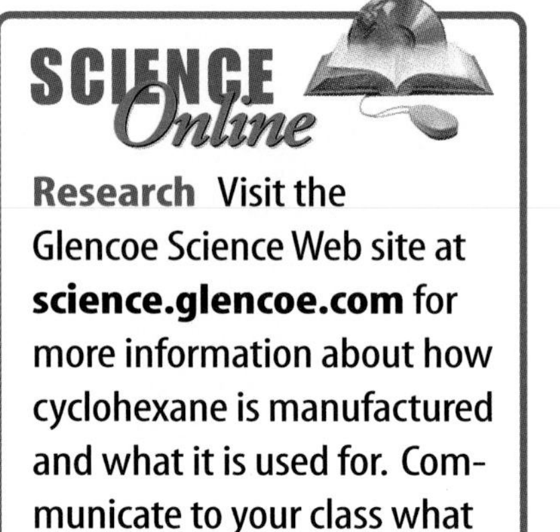

Research Visit the Glencoe Science Web site at **science.glencoe.com** for more information about how cyclohexane is manufactured and what it is used for. Communicate to your class what you learn.

Hydrocarbons in Rings You might be thinking that all hydrocarbons are chains of carbon atoms with two ends. Some molecules contain rings. You can see the structures of two different molecules in **Figure 9.** The carbon atoms of hexane bond together to form a closed ring containing six carbons. Each carbon atom still has four bonds. The prefix *cyclo-* in their names tells you that the molecules are cyclic, or ring shaped.

Ring structures are not uncommon in chemical compounds. Many natural substances such as sucrose, glucose, and fructose are ring structures. Ring structures can contain one or more double bonds.

Reading Check ***What does the prefix*** **cyclo-** ***tell you about a molecule?***

Section 1 Assessment

1. What is an organic compound? Give an example of an item that contains an organic compound.
2. Explain the difference between a saturated hydrocarbon and an unsaturated hydrocarbon and give an example of each.
3. Based on the structure of the carbon atom, explain the large number of compounds that can be formed by carbon.
4. Explain what the term *isomer* means and draw all of the possible isomers for pentene, C_5H_{10}.
5. **Think Critically** Are propane and cyclopropane isomers? Draw their structures. Use their structures and formulas to explain your answer.

Skill Builder Activities

6. **Making and Using Graphs** Make a graph using the information in **Table 1.** For each compound, plot the number of carbon atoms on the *x*-axis and the number of hydrogen atoms on the *y*-axis. Use your graph to predict the formula for the saturated hydrocarbon that has 11 carbon atoms. **For more help, refer to the Science Skill Handbook.**
7. **Solving One-Step Equations** The general formula for saturated hydrocarbons is C_nH_{2n+2} where *n* can be any whole number except zero. Use the general formula to determine the chemical formula for a saturated hydrocarbon with 25 carbon atoms. **For more help, refer to the Math Skill Handbook.**

SECTION 2

Other Organic Compounds

Substituted Hydrocarbons

Suppose you pack an apple in your lunch every day. One day, you have no apples, so you substitute a pear. When you eat your lunch, you'll notice a difference in the taste and texture of your fruit. Chemists make substitutions, too. They change hydrocarbons to make compounds called substituted hydrocarbons. To make a substituted hydrocarbon, one or more hydrogen atoms are replaced by atoms such as halogens or by groups of atoms. Such changes result in compounds with chemical properties different from the original hydrocarbon. When one or more chlorine atoms are added to methane in place of hydrogens, new compounds are formed. **Figure 10** shows the four possible compounds formed by substituting chlorine atoms.

As You Read

What **You'll Learn**

- **Describe** how new compounds are formed by substituting hydrocarbons.
- **Identify** the classes of compounds that result from substitution.

Vocabulary

hydroxyl group
carboxyl group
amino group
amino acid

Why **It's Important**

Many natural and manufactured organic compounds are formed by replacing hydrogen with other atoms.

Figure 10
Chlorine can replace hydrogen atoms in methane.

A **Chloromethane contains a single chlorine atom.**

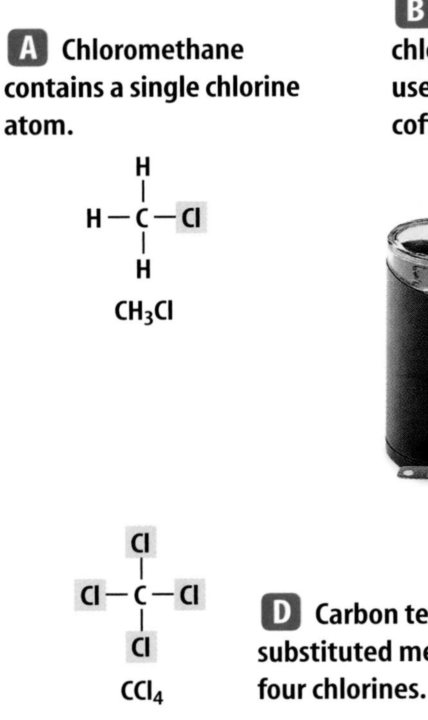

B **Dichloromethane contains two chlorine atoms. Dichloromethane is used in manufacturing decaffeinated coffee.**

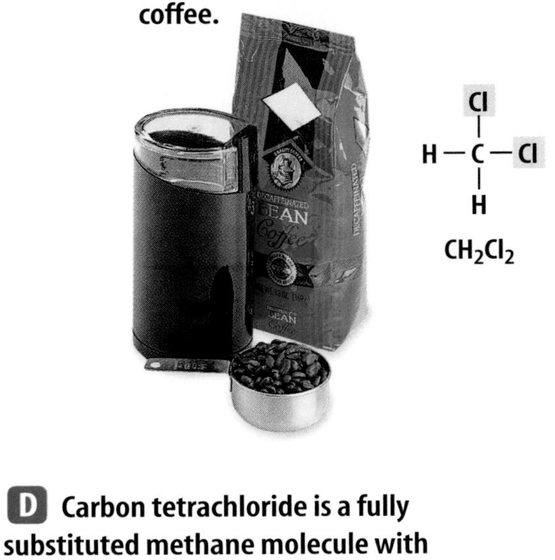

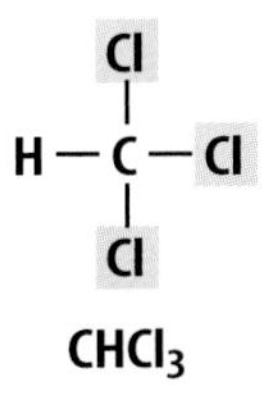

C **Trichloromethane, or chloroform, has three chlorine atoms. Chloroform is used as a veterinary anesthetic.**

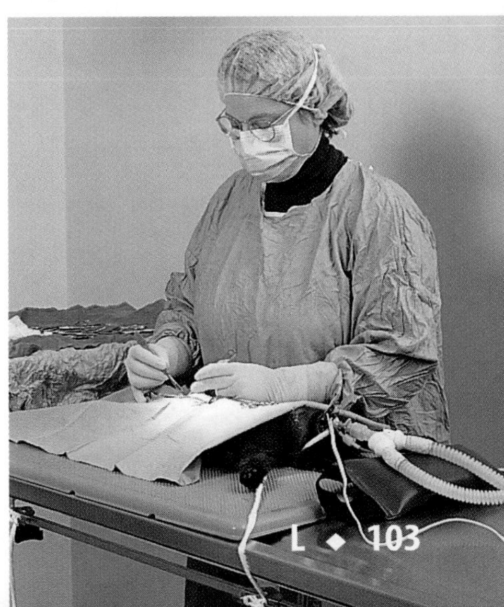

D **Carbon tetrachloride is a fully substituted methane molecule with four chlorines.**

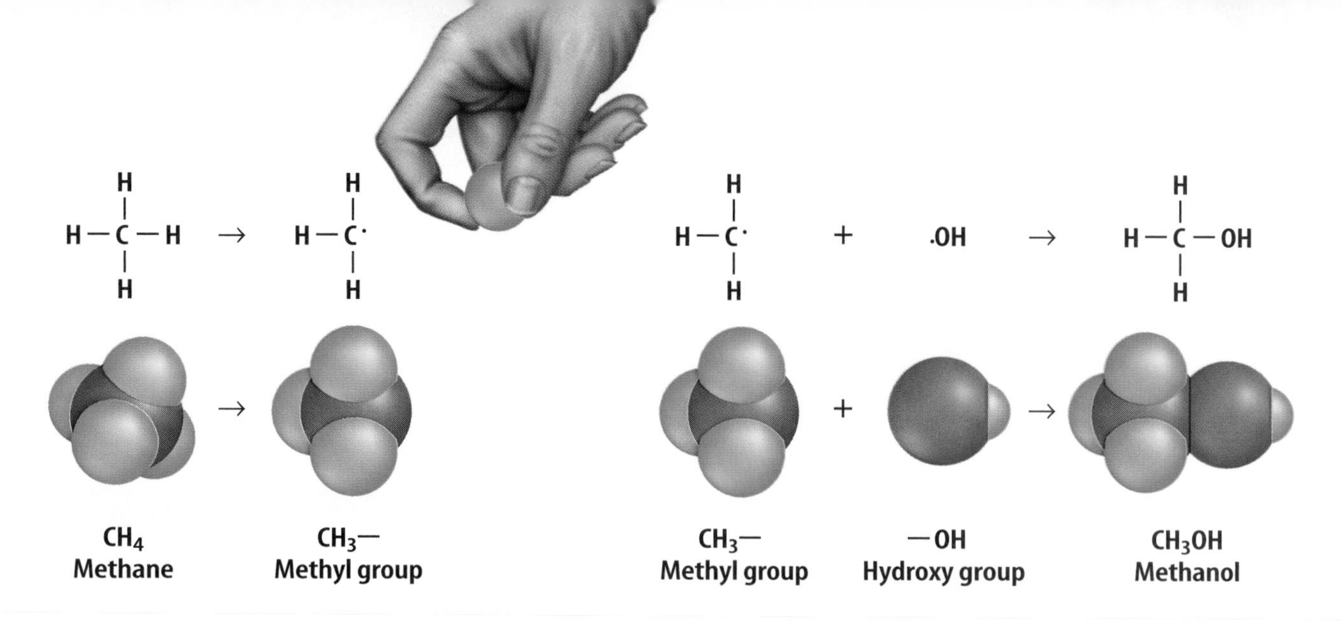

Figure 11
After the methane molecule loses one of its hydrogen atoms, it has an extra electron to share, as does the hydroxyl group.
What kind of bond do they form?

Alcohols Groups of atoms also can be added to hydrocarbons to make different compounds. The **hydroxyl group** (hi DROK sul) is made up of an oxygen atom and a hydrogen atom joined by a covalent bond. A hydroxyl group is represented by the formula —OH. An alcohol is formed when a hydroxyl group replaces a hydrogen atom in a hydrocarbon. **Figure 11** shows the formation of the alcohol methanol. A hydrogen atom in the methane molecule is replaced by a hydroxyl group.

Reading Check *What does the formula —OH represent?*

Larger alcohol molecules are formed by adding more carbon atoms to the chain. Ethanol is an alcohol produced naturally when sugar in corn, grains, and fruit ferments. It is a combination of ethane, which contains two carbon atoms, and an —OH group. Its formula is C_2H_6O. Isopropyl alcohol forms when the hydroxyl group is substituted for a hydrogen atom on the middle carbon of propane rather than one of the end carbons. **Table 2** lists three alcohols with their structures and uses. You've probably used isopropyl alcohol to disinfect injuries. Did you know that ethanol can be added to gasoline and used as a fuel for your car?

Table 2 Common Alcohols

Uses	Methanol (CH_3OH: H—C—OH with H above and below C)	Ethanol (H—C—C—OH, each C with H above and below)	Isopropyl Alcohol (H—C—C—C—H, middle C bearing OH below)
Fuel	yes	yes	no
Cleaner	yes	yes	yes
Disinfectant	no	yes	yes
Manufacturing chemicals	yes	yes	yes

Carboxylic Acids Have you ever tasted vinegar? Vinegar is a solution of acetic acid and water. You can think of acetic acid as the hydrocarbon methane with a carboxyl group substituted for a hydrogen. A **carboxyl group** (car BOK sul) consists of a carbon atom, two oxygen atoms, and a hydrogen atom. Its formula is $-COOH$. When a carboxyl group is substituted in a hydrocarbon, the substance formed is called a carboxylic acid. The simplest carboxylic acid is formic acid. Formic acid consists of a single hydrogen atom and a carboxyl group. You can see the structures of formic acid and acetic acid in **Figure 12.**

You probably can guess that many other carboxylic acids are formed from longer hydrocarbons. Many carboxylic acids occur in foods. Citric acid is found in citrus fruits such as oranges and grapefruit. Lactic acid is present in sour milk. Acetic acid dissolved in water—vinegar—often is used in salad dressings.

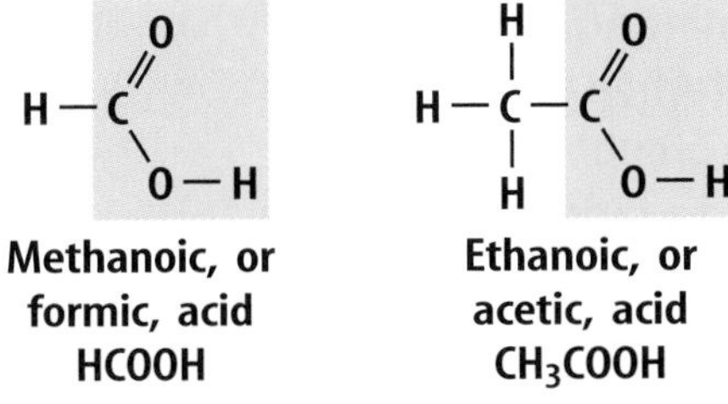

Figure 12
Crematogaster **ants make the simplest carboxylic acid, formic acid. Notice the structure of the $-COOH$ group.** *How do the structures of formic acid and acetic acid differ?*

Amines Amines are a group of substituted hydrocarbons formed when an amino group replaces a hydrogen atom. An **amino** (uh ME noh) **group** is a nitrogen atom joined by a covalent bond to two hydrogen atoms. It has the formula $-NH_2$. Methylamine, shown in **Figure 13,** is formed when one of the hydrogens in methane is replaced with an amino group. A more complex amine that you might have experienced is the novocaine your dentist uses to numb your mouth during dental work. Amino groups are important because they are a part of many biological compounds that are essential for life.

Figure 13
Complex amines account for the strong smells of cheeses such as these.

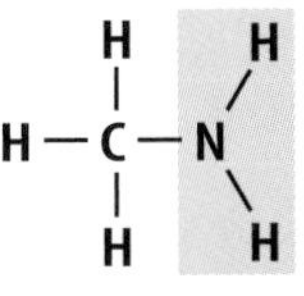

Amino Acids You have seen that a carbon group can be substituted onto one end of a chain to make a new molecule. It's also possible to substitute groups on both ends of a chain and even to replace hydrogen atoms bonded to carbon atoms in the middle of a chain. When both an amino group ($-NH_2$) and a carboxyl acid group ($-COOH$) replace hydrogens on the same carbon atom in a molecule, a type of compound known as an **amino acid** is formed. Amino acids are essential for human life.

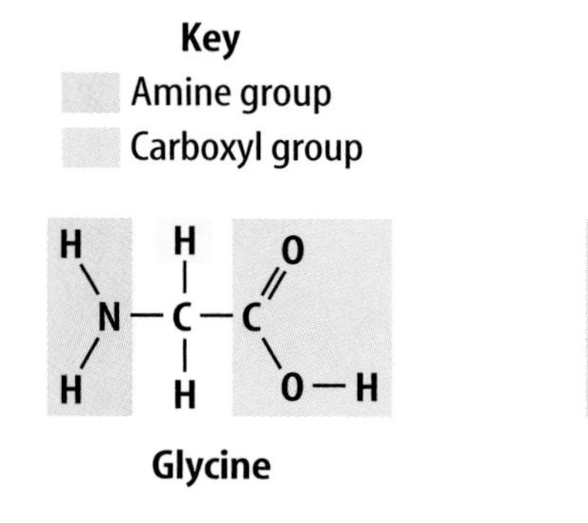

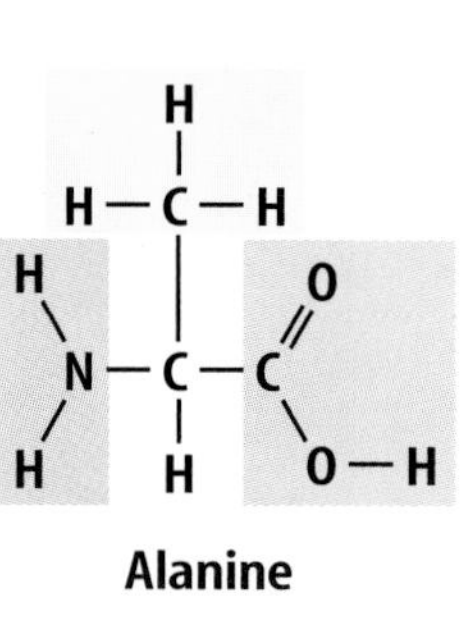

Figure 14
The 20 amino acids each contain a central carbon atom bonded to an amine group, a hydrogen atom, and a carboxyl group. The fourth bond, shown in yellow, is different for each amino acid.

The Building Blocks of Protein

Amino acids are the building blocks of proteins, which are an important class of biological molecules needed by living cells. Twenty different amino acids bond in different combinations to form the variety of proteins that are needed in the human body. Glycine and alanine are shown in **Figure 14.** Glycine is the simplest amino acid. It is a methane molecule in which one hydrogen atom has been replaced by an amine group and another has been replaced by a carboxyl group. The other 19 amino acids are formed by replacing the yellow highlighted hydrogen atom with different groups. For example, in alanine, one hydrogen atom is replaced by a methyl ($-CH_3$) group.

Some amino acids such as glycine and alanine are manufactured within the human body. Others are obtained by eating protein-rich foods.

Reading Check *What are the building blocks of protein?*

Section 2 Assessment

1. The nonstick coating found on some pots and pans is made from tetrafluoroethylene, a substituted hydrocarbon in which all four of the hydrogen atoms of ethylene are replaced by fluorine. Draw the structural formula for this molecule.
2. In what way is an amino acid different from a carboxylic acid?
3. How do the 20 amino acids differ from each other?
4. If the hydroxyl group replaces a hydrogen in a hydrocarbon, what type of compound is formed?
5. **Think Critically** The formula for one compound that produces the odor in skunk spray is $CH_3CH_2CH_2CH_2SH$. Draw and examine the structural formula for this compound. Does it fit the definition of a substituted hydrocarbon? Explain.

Skill Builder Activities

6. **Predicting** Compounds in which hydrogen atoms in methane have been replaced by chlorine and fluorine atoms are known as chlorofluorocarbons (CFCs). They were used in aerosol cans and as refrigerants until it was found that they can damage Earth's ozone layer. Draw the structure of three CFC and predict what their chemical names might be. **For more help, refer to the Science Skill Handbook.**
7. **Using a Word Processor** Use the table function in a word processing program to make a table listing the classes of substituted hydrocarbons in this section: *halogen-substituted hydrocarbons, alcohols, carboxylic acids, amines,* and *amino acids.* List the substituted group(s) for each class and give the name and formula of a molecule that belongs in each class. **For more help, refer to the Technology Skill Handbook.**

Activity

Conversion of Alcohols

Have you ever wondered how chemists change one substance into another? In this activity, you will change an alcohol into an acid.

What You'll Investigate

What changes occur when ethanol is exposed to conditions like those produced by exposure to air and bacteria?

Materials

test tube and stopper
test-tube rack
potassium permanganate solution (1 mL)
sodium hydroxide solution (1 mL)
ethanol (3 drops)
pH test paper
graduated cylinder

Goals

- **Observe** a chemical change in an alcohol.
- **Infer** the product of the chemical change.

Safety Precautions

WARNING: *Handle these chemicals with care. Immediately flush any spills with water and call your teacher. Keep your hands away from your face. Wash your hands after completing the experiment.*

Procedure

1. **Measure** 1 mL of potassium permanganate solution and pour it into a test tube. Carefully measure 1 mL of sodium hydroxide solution and add it to the test tube.

Alcohol Conversion	
Procedure Step	Observations
Step 2	
Step 3	
Step 4	
Step 5	

2. With your teacher's help, dip a piece of pH paper into the mixture in the test tube. Record the result in your Science Journal.
3. Add three drops of ethanol to the test tube. Put a stopper on the test tube and gently shake it for one minute. Record any changes.
4. Place the test tube in a test-tube rack. Observe and record any changes you notice during the next five minutes.
5. Test the sample with pH paper again. Record what you observe.
6. Your teacher will dispose of the solutions.

Conclude and Apply

1. Did a chemical reaction take place? What leads you to infer this?
2. Alcohols can undergo a chemical reaction to form carboxylic acids in the presence of potassium permanganate. If the alcohol used is ethanol, what would you predict to be the chemical formula of the acid produced?

Communicating Your Data

Compare your conclusions with other students in your class. **For more help, refer to the** Science Skill Handbook.

Biological Compounds

As You Read

What You'll Learn

- **Describe** how large organic molecules are made.
- **Explain** the roles of organic molecules in the body.
- **Explain** why eating a balanced diet is important for good health.

Vocabulary

polymer
protein
carbohydrate
lipid

Why It's Important

Organic molecules are important to your body processes.

What's a polymer?

Now that you know about some simple organic molecules, you can begin to learn about more complex molecules. One type of complex molecule is called a polymer (PAH luh mur). A **polymer** is a molecule made up of many small organic molecules linked together with covalent bonds to form a long chain. The small, organic molecules that link together to form polymers are called monomers. Polymers can be produced by living organisms or can be made in a laboratory. Polymers produced by living organisms are called natural polymers. Polymers made in a laboratory are called synthetic polymers.

Reading Check *What is a polymer, and how does it resemble a chain?*

To picture what polymers are, it is helpful to start with small synthetic polymers. You use such polymers every day. Plastics, synthetic fabrics, and nonstick surfaces on cookware are polymers. The unsaturated hydrocarbon ethylene, C_2H_4, are the monomers of a common polymer used often in plastic bags. The ethylene monomers are joined in a chemical reaction called polymerization (puh lih muh ruh ZAY shun). As you can see in **Figure 15,** one of the double bonds breaks in each ethylene molecule. The two carbon atoms then form new bonds with carbon atoms in other ethylene molecules. This process is repeated many times and results in a much larger molecule called polyethylene. A polyethylene molecule can contain 10,000 ethylene units.

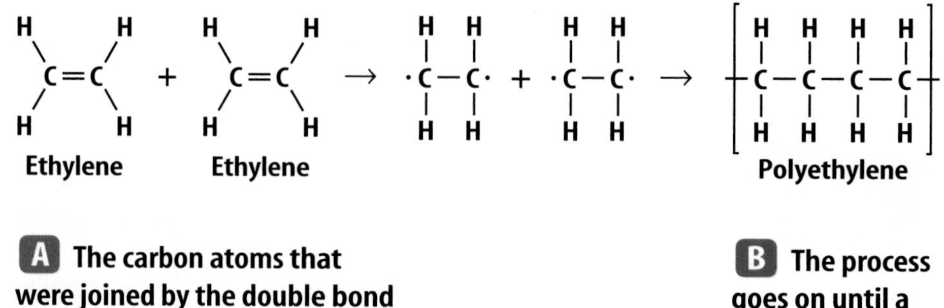

Figure 15
Small molecules called monomers link into long chains to form polymers.

A **The carbon atoms that were joined by the double bond each have an electron to share with another carbon in another molecule of ethylene.**

B **The process goes on until a long molecule is formed.**

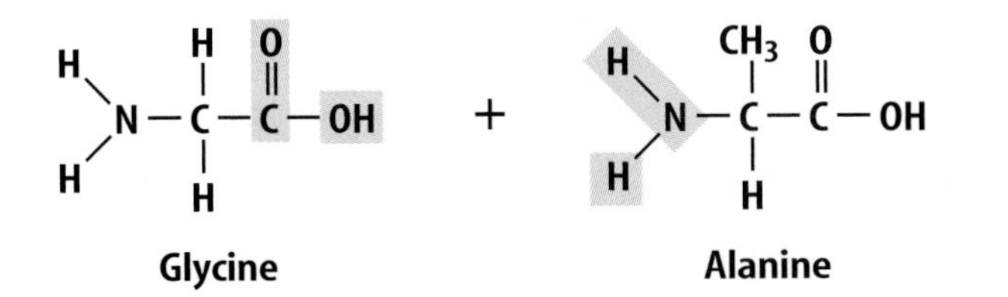

Proteins Are Polymers

You've probably heard about proteins when you've been urged to eat healthful foods. A **protein** is a polymer that consists of a chain of individual amino acids linked together. Your body cannot function properly without them. Proteins in the form of enzymes serve as catalysts and speed up chemical reactions in cells. Some proteins make up the structural materials in ligaments, tendons, muscles, cartilage, hair, and fingernails. Hemoglobin, which carries oxygen through the blood, is a protein polymer, and all body cells contain proteins.

The various functions in your body are performed by different proteins. Your body makes many of these proteins by assembling 20 amino acids in different ways. Eight of the amino acids that are needed to make proteins cannot be produced by your body. These amino acids, which are called essential amino acids, must come from the food you eat. That's why you need to eat a diet containing protein-rich foods, like those in **Table 3.**

The process by which your body converts amino acids to proteins is shown in **Figure 16.** In this reaction, the amino group of the amino acid alanine forms a bond with the carboxyl group of the amino acid glycine, and a molecule of water is released. Each end of this new molecule can form similar bonds with another amino acid. The process continues in this way until the amino acid chain, or protein, is complete.

Reading Check *How is an amino acid converted to protein?*

Table 3 Protein Content (Approximate)

Foods	Protein Content (g)
Cheese Pizza (2 slices)	30
Cheeseburger	30
Whole Milk (240 mL)	8
Peanut Butter (30 g)	8
Kidney Beans (127 g)	8

Figure 16
Both ends of an amino acid can link with another amino acid.
What molecule is released in the process?

Summing Up Protein

Procedure

1. Make a list of the foods you ate during the last 24 h.
2. Use the data your teacher gives you to find the total number of grams of protein in your diet for the day. Multiply the grams of protein in one serving of food by the number of units of food you ate. The recommended daily allowance (RDA) of protein for girls 11 to 14 years old is 46 g per day. For boys 11 to 14 years old, the RDA is 48 g per day.

Analysis

1. Was your total greater or less than the RDA?
2. Which of the foods you ate supplied the largest amount of protein? What percent of the total grams of protein did that food supply?

Figure 17
These foods contain a high concentration of carbohydrates.

Carbohydrates

The day before a race, athletes often eat large amounts of foods containing carbohydrates like the ones in **Figure 17.** What's in pasta and other foods like bread and fruit that gives the body a lot of energy? These foods contain sugars and starches, which are members of the family of organic compounds called carbohydrates. A **carbohydrate** is an organic compound that contains only carbon, hydrogen, and oxygen, usually in a ratio of two hydrogen atoms to one oxygen atom. In the body, carbohydrates are broken down into simple sugars that the body can use for energy. The different types of carbohydrates are divided into groups—sugars, starches, and cellulose.

Table 4 below gives some approximate carbohydrate content for some of the common food groups.

Health INTEGRATION

Carbohydrates are the best energy sources for an athelete. In an athletic event, carbohydrates are the fuel source for the body. Carbohydrate-rich foods include breads, cereals, pastas, starchy vegetables such as corn and potatoes, and dried beans and peas. Fruits are also a good source of carbohydrates. It is important to eat a variety of these foods to maintain good health.

Table 4 Carbohydrates in Foods (Approximate)

Foods	Carbohydrate Content (g)
Bread (1 slice)	15
Fruit Serving (120 mL)	10–15
Milk (240 mL)	12
Starchy Vegetable (120 mL)	15
Rice (120 mL)	15

Figure 18
Glucose and fructose are simple six-carbon carbohydrates found in many fresh and packaged foods. Glucose and fructose are isomers. *Can you explain why?*

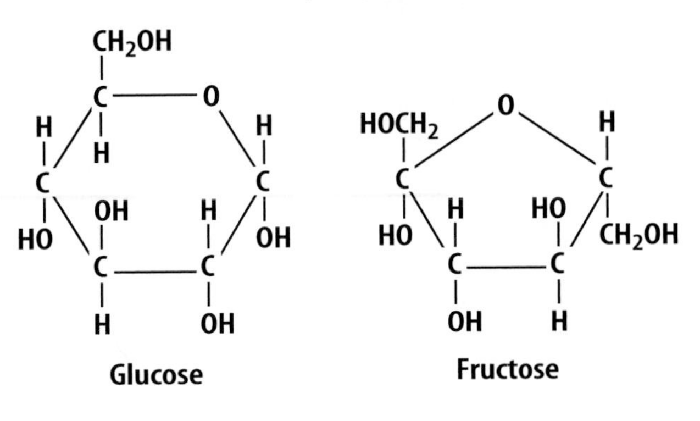

Sugars If you like chocolate-chip cookies or ice cream, then you're familiar with sugars. They are the substances that make fresh fruits and candy. Simple sugars are carbohydrates containing five, six, or seven carbon atoms arranged in a ring. The structures of glucose and fructose, two common simple sugars, are shown in **Figure 18.** Glucose forms a six-carbon ring. It is found in many naturally sweet foods, such as grapes and bananas. Fructose is the sweet substance found in ripe fruit and honey. It often is found in corn syrup and added to many foods as a sweetener. The sugar you probably have in your sugar bowl or use in baking a cake is sucrose. Sucrose, shown in **Figure 19,** is a combination of the two simple sugars glucose and fructose. In the body, sucrose cannot move through cell membranes. It must be broken down into glucose and fructose to enter cells. Inside the cells, these simple sugars are broken down further, releasing energy for cell functions.

Starches Starches are large carbohydrates that exist naturally in grains such as rice, wheat, corn, potatoes, lima beans, and peas. Starches are polymers of glucose monomers in which hundreds or even thousands of glucose molecules are joined together. Because each sugar molecule releases energy when it is broken down, starches are sources of large amounts of energy.

Figure 19
Sucrose is a molecule of glucose combined with a molecule of fructose. *What small molecule must be added to sucrose when it separates to form the two six-carbon sugars?*

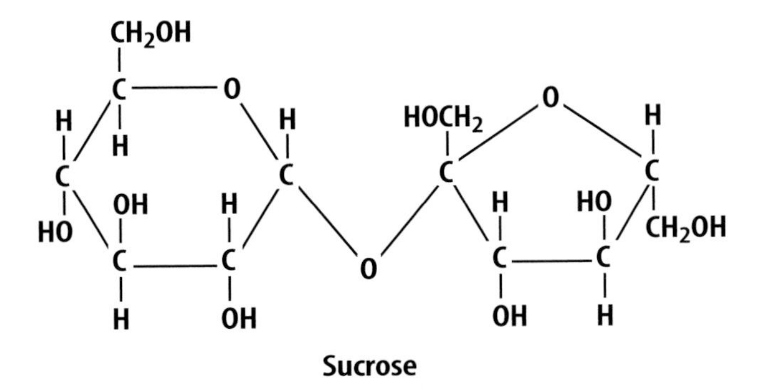

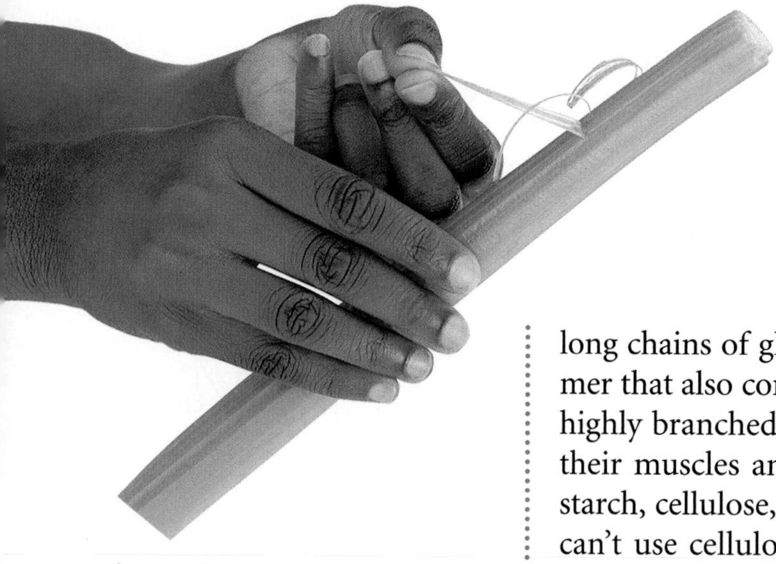

Other Glucose Polymers Two other important polymers that are made up of glucose molecules are cellulose and glycogen. Cellulose makes up the long, stiff fibers found in the walls of plant cells, like the strands that pull off the celery stalk in **Figure 20.** It is a polymer that consists of long chains of glucose units linked together. Glycogen is a polymer that also contains chains of glucose units, but the chains are highly branched. Animals make glycogen and store it mainly in their muscles and liver as a ready source of glucose. Although starch, cellulose, and glycogen are polymers of glucose, humans can't use cellulose as a source of energy. The human digestive system can't convert cellulose into sugars. Grazing animals, such as cows, have special digestive systems that allow them to break down cellulose into sugars.

Figure 20
Your body cannot chemically break down the long cellulose fibers in celery, but it needs fiber to function properly.

Reading Check *How do the location and structure of glycogen and cellulose differ?*

Problem-Solving Activity

Which foods are best for quick energy?

Foods that are high in carbohydrates are sources of energy. But which foods are best?

Identifying the Problem

The chart shows some common foods and their carbohydrate count. Look at the chart to see the differences in how much energy they might provide, given their carbohydrate count.

Practice Problems

1. From the chart, create a high-energy meal with the most carbohydrates. Include one main dish, one side dish, and one drink. Create another meal that contains a maximum of 60 g of carbohydrates.
2. Meat and many vegetables have only trace amounts of carbohydrates. Considering this, how many grams of carbohydrates would a meal of hamburger, fries, and a drink contain?

High-Carbohydrate Foods

Main Dish		Side Dish		Drink	
two slices white bread	28 g	fudge brownie	25 g	orange juice	29 g
macaroni and cheese	48 g	slice of yellow cake	53 g	cola	42 g
two pancakes	48 g	bagel	30 g	root beer	44 g
one toaster pastry	39 g	blueberry muffin	27 g	lemon-lime soda	39 g
hamburger bun	21 g	cake donut	30 g	hot cocoa	25 g
sweetened flakes	35 g	glazed donut	22 g	apple juice	30 g
toasted oats	23 g	mashed potatoes	26 g	lemonade	28 g
bran flakes with raisins	45 g	stuffing	19 g	milk	12 g
white rice	50 g	roll	15 g	chocolate milk	28 g
spaghetti	39 g	20 french fries	36 g	sports drink	14 g

Procedure

1. Collect all the materials and equipment you will need.
2. Obtain 10 mL of the starch solution from your teacher.
3. Add the starch solution to 200 mL of water in a 250-mL beaker. Stir.
4. Add four drops of iodine solution to the beaker to make a dark-blue indicator solution.
5. Obtain your teacher's approval of your indicator solution before proceeding.
6. **Measure** and place 5 mL of the indicator solution in a clean test tube.
7. Obtain 5 mL of vitamin-C solution from your teacher.
8. Using a clean medicine dropper, add one drop of the vitamin-C solution to the test tube. Stir. Continue adding drops and stirring until you notice a color change. Place a piece of white paper behind the test tube to show the color clearly. Record the number of drops added and any observations.
9. Using a clean test tube and dropper for each test, repeat steps 6 and 8, replacing the vitamin-C solution with other liquids and solids. Add drops of liquid foods or juices until a color change is noted or until you have added about 40 drops of the liquid. Mash solid foods such as onion and potato. Add about 1 g of the food directly to the test tube and stir. Test at least four liquids and four solids.
10. **Record** the amount of each food added and observations in a table.

Conclude and Apply

1. What indicates a positive test for vitamin C? How do you know?
2. **Describe** a negative test for vitamin C.
3. Which foods tested positive for vitamin C? Which foods, if any, tested negative for vitamin C?
4. Which foods might you include in your diet to make sure you get vitamin C every day? Could a vitamin-C tablet take the place of these foods? Explain.

Communicating Your Data

Compare your results with other class members. Were your results consistent? Make a record of the foods you eat for two days. Does your diet contain the minimum RDA of vitamin C?

TIME SCIENCE AND Society

SCIENCE ISSUES THAT AFFECT YOU!

From Plants to Medicine

Wild plants help save lives

Look carefully at those plants growing in your backyard or neighborhood. With help from scientists, they could save a life. Many of the medicines that doctors prescribe were first developed from plants. For example, aspirin was extracted from the bark of a willow tree. A cancer medication was extracted from the bark of the Pacific yew tree. Aspirin and and the cancer medication are now made synthetically—their carbon structures are duplicated in the lab and factory.

Throughout history, and in all parts of the world, traditonal healers have used different parts of plants and flowers to help treat people. Certain kinds of plants have been mashed up and applied to the body to heal burns and sores, or have been swallowed or chewed to help people with illnesses.

Promising cancer medications are made from the bark of the Pacific yew tree.

Workers at a hospital in India squeeze juice from brahmi, or water hyssop. Mixed with other plants and melted butter, the extract is used in many parts of the world to treat mental illnesses.

16. Making and Using Graphs The graph shows the boiling points of some saturated hydrocarbons. How does boiling point vary with the number of carbon atoms? What do you predict would be the approximate boiling point of hexane?

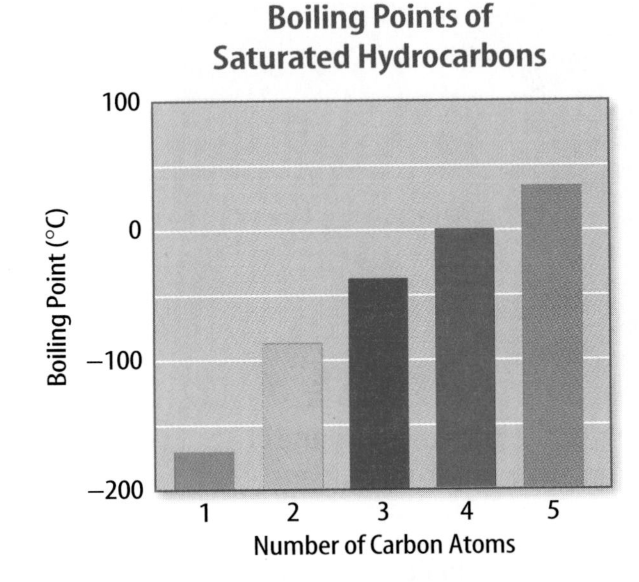

17. Hypothesizing PKU is a genetic disorder that can lead to brain damage. People with this disorder cannot process one of the amino acids. Luckily, damage can be prevented by a proper diet. How is this possible?

Performance Assessment

18. Scientific Drawing Research an amino acid that was not mentioned in the chapter. Draw its structural formula and highlight the portion that substitutes for a hydrogen atom.

Technology

Go to the Glencoe Science Web site at **science.glencoe.com** or use the **Glencoe Science CD-ROM** for additional chapter assessment.

THE PRINCETON REVIEW Test Practice

The box below shows the structural formula of propane, which is used as fuel.

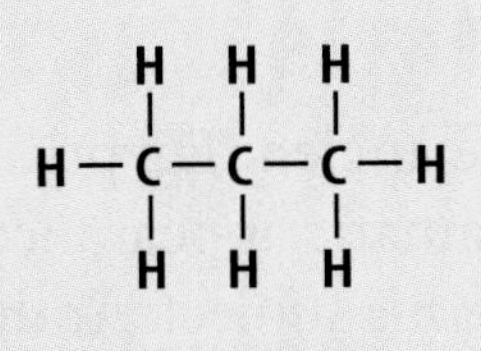

Study the picture and answer the following questions.

1. The compound shown above is a _____.

A) saturated hydrocarbon
B) unsaturated hydrocarbon
C) saturated organic acid
D) unsaturated organic acid

2. The molecular formula of the compound shown above is _____.

F) CH
G) C_3H_8
H) C_3H_8O
J) C_3H_{10}

3. The next compound in this series is _____.

A) methane
B) ethane
C) butane
D) hexane

4. If an −OH group is substituted for a hydrogen, this compound is a(n) _____.

F) alcohol
G) carboxylic acid
H) amine
J) amino acid

Standardized Test Practice

Reading Comprehension

Read the passage. Then read each question that follows the passage. Decide which is the best answer to each question.

Test-Taking Tip Remember that clues and information can always be found in the passage.

The Eyes Have It

Notice in the ancient Egyptian artwork on the right the dramatic appearance of the dark-rimmed eyes of this subject. Anthropologists know from the records of ancient scribes that the Egyptians achieved this effect in real life by using cosmetics. However, until recently, they were unaware of the sophisticated chemistry that went into the ancient Egyptians' creation of these early cosmetics.

The Louvre Museum in Paris has an extensive collection of ancient Egyptian makeup containers. Amazingly, many of them still retain remnants of their original contents. When the technology to analyze the remains of the Egyptians' creams and powders became available, French scientists went to work. They found that the cosmetics contained several lead-based chemicals. Two of them, lead sulfide (PbS) and lead carbonate ($PbCO_3$), are common in nature. However, two of the others found in the makeup are not. $PbOHCl$ and $Pb_2Cl_2CO_3$ are formed only when lead minerals oxidize in a combination of chlorinated and carbonated water. Scientists believe that the Egyptians created these compounds artificially, using a sophisticated process called "wet chemistry" to synthesize their molecules.

Scientists already had found written records indicating that the ancient Greeks and Romans used wet chemistry. However, the Egyptian discovery pushes the date of the first known use of this technique back to 2000 B.C., more than 1,000 years before the birth of Cleopatra.

This statue reflects regular use of cosmetics by ancient Egyptians.

1. Scientists believe that Egyptians artificially created two of the compounds found in the ancient cosmetics because _____.

A) scientists had never seen these compounds before

B) they are not common in nature

C) the process used to make them was simple

D) only artificially created compounds could survive such a long time

2. Historians knew that ancient Greeks and Romans used "wet chemistry" because _____.

F) they had found samples of cosmetics for testing

G) they saw evidence of the use of cosmetics in their artwork

H) surviving text from the period indicates the use of "wet chemisty"

J) the Greek and Roman civilizations followed the Egyptian model

Standardized Test Practice

THE PRINCETON REVIEW

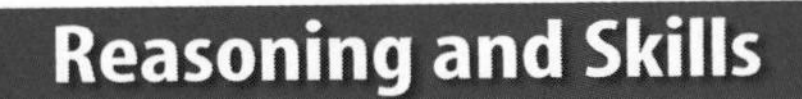

Reasoning and Skills

Read each question and choose the best answer.

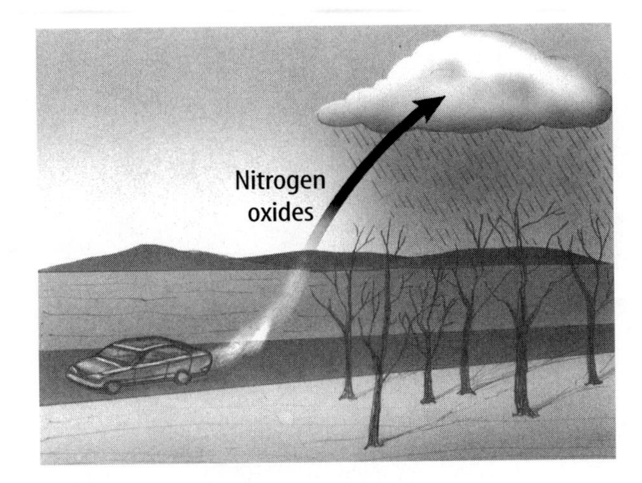

1. This picture indicates that the tree is probably losing its leaves because of _____.

A) soil erosion
B) acid rain
C) nutrient-poor soil
D) overwatering

Test-Taking Tip The best answer is always supported by any information in a chart or graphic.

2. What is the source of the nitrogen oxides?

F) the soil
G) the air
H) the car exhaust
J) the cloud

Common Bases and Applications	
Bases	**Product**
Sodium hydroxide	Drain cleaner
Magnesium hydroxide	Laxative
Ammonium hydroxide	Window cleaner
Aluminum hydroxide	Deodorant

3. According to the information on the table at the bottom of the first column, ammonium hydroxide is used in the manufacturing of _____.

A) window cleaner
B) laxative
C) drain cleaner
D) deodorant

Test-Taking Tip Always consider every answer choice as you work through a question.

4. A balanced chemical equation has the same number of atoms of each element on both sides of the equation. Which of the following is a balanced equation?

F) $2Al + 3Ag_2S \rightarrow Al_2S_3 + Ag$
G) $2H_2CO_3 \rightarrow H_2O + CO_2$
H) $Fe + O_2 \rightarrow Fe_2O_3$
J) $2Ag + H_2S \rightarrow Ag_2S + H_2$

Test-Taking Tip Double-check that you have identified the correct answer by making sure that the other answer choices are not balanced equations.

Consider this question carefully before writing your answer on a separate sheet of paper.

5. Explain the difference between a group of elements and a period—both found on the periodic table of elements. Give an example of a group found on the periodic table.

Test-Taking Tip Recall how the periodic table is designed.

Student Resources

CONTENTS

Field Guides 128

Skill Handbooks 132

Reference Handbook 157

It's early morning in the kitchen, and, whether you know it or not, chemistry is occurring all around you. Breakfast—with its wake-up sights and smells—is almost ready. Butter, syrup, freshly squeezed orange juice, hot tea, and yogurt with strawberries wait on the counter. Eggs and pancakes sizzle on the griddle. Slices of bread are toasting. Some foods are liquids. Others are solids. Most are mixtures. Some of these are undergoing changes while you watch. Using this field guide, you can identify the different types of mixtures you drink and eat, and the chemical and physical changes that occur as foods are prepared.

How are mixtures classified?

Mixtures contain two or more substances that have not combined to form a new substance. The proportions of the substances that make up a mixture can vary. Mixtures are classified as homogeneous or heterogeneous.

You cannot see the separate substances in a homogeneous mixture no matter how closely you look. Cranberry juice is a homogenous mixture. You can easily identify the separate substances that are in a heterogeneous mixture. Breakfast cereals are heterogeneous mixtures.

Kitchen Chemistry

Homogeneous Mixtures

Homogeneous mixtures can be solids, liquids, or gases. Stainless steel, for example, is a solid mixture of iron, carbon, and chromium. You might have cookware, containers, and utensils made of stainless steel. Also found in abundance in your kitchen is a familiar mixture of gases, primarily nitrogen, oxygen, and argon. This mixture is the air you breathe. Much of the chemistry in your kitchen occurs in solutions. Solutions are homogeneous mixtures, therefore you cannot see their different parts. Tea and syrup are solutions of solids that are dissolved in liquids.

Tea and syrup

Stainless steel is a homogeneous mixture.

Field Activity

Use this field guide to help identify the mixtures and changes in your kitchen. Observe the preparation of a few meals. In your Science Journal, record the meal being prepared and a description of the types of mixtures, chemical changes, and physical changes you observe.

Heterogeneous Mixtures

You can see the different parts of a heterogeneous mixture. Familiar heterogeneous mixtures tend to be solids or solids and liquids. For example, the strawberries are visible in the bowl of yogurt and so are the blueberries in the muffin.

If you have left butter or cooking oil heating too long in a frying pan, you know that the smoke that rises from the pan is visible in the air. This mistake created a heterogeneous mixture.

Blueberry muffins and yogurt with strawberries are heterogeneous mixtures.

How are changes classified?

A change to a substance can be classified as chemical or physical. During a chemical change, one or more new substances are formed. When a physical change occurs, the identity of the substance remains the same.

Chemical Changes in the Kitchen

You can recognize a chemical change if one or more of the following occurs: the substance changes color, the substance produces a new odor, the substance absorbs or releases heat or light; or the substance releases a gas.

Browning

Browning is a chemical change that occurs when sugars and proteins in foods form new flavors and smells. It produces the barbecue flavors of foods that are cooked on a grill and the caramelized flavor of a roasted marshmallow.

Grilled hamburgers and vegetables are examples of chemical changes.

A marshmallow is browning on a skewer.

Protein Denaturation

A protein in its natural state is a long chain of chemical units. After it has formed, this chain folds into a specific shape determined by attractions, or weak bonds, between chemical units in that protein. Denaturation involves breaking the weak bonds in a protein and changing its shape.

The proteins in a raw egg are folded into balls, sheets, and coils. Heating the egg breaks some of the weak bonds that hold the proteins in these tight shapes. As cooking continues, the proteins unravel and begin forming weak bonds with other proteins. This causes the egg to solidify.

Eggs that are cooked sunny side up are an example of protein denaturation.

Bubbling pancake batter shows gas production.

Gas Production

The new substances that are produced during a chemical change are sometimes gases. Gas production occurs in the preparation of some foods.

The bubbles you see in pancake batter as it cooks are caused by a chemical change in the batter. Baking powder is a mixture of baking soda, $NaHCO_3$ (sodium bicarbonate), and an acidic substance. When water is added to baking powder, the acidic solution that forms reacts with baking soda to make carbon dioxide gas. As a pancake cooks, the bubbles of carbon dioxide rise through the batter, leaving spaces that fluff out the pancake making it light in texture.

Physical Changes in the Kitchen

You can recognize a physical change when the substance changes shape, size, or state. For example, water can become ice, or a chocolate bar can melt.

Melting

Melting is the physical change in which a solid becomes a liquid. The pat of butter is changing from a solid to a liquid. Melting occurs because heat from the warm toast weakens the attractions between the molecules of butter. The cheese in a grilled-cheese sandwich is another good example.

Melting butter is an example of a solid becoming a liquid.

Freezing

Freezing is the physical change in which a liquid becomes a solid. Chemical substances have a freezing point—the temperature at which this change occurs. Most of the water in a liquid ice-cream mixture freezes into small ice crystals, and air bubbles give the solid mixture its smooth, creamy texture. Water frozen in the form of ice cubes is used to chill beverages. Freezing also is used to preserve a wide variety of foods.

Making ice cream demonstrates how a liquid becomes a solid.

Boiling

Boiling is the physical change in which a liquid becomes a gas. Popcorn kernels contain 11 percent to 14 percent water. When it is heated, that water changes to steam. Because steam takes up many times the volume of liquid water, it creates enough pressure to burst the kernels. Some cereals, vegetables, and other foods can be cooked in boiling water or in the steam that is produced when water boils.

Popcorn is produced by boiling.

Organizing Information

As you study science, you will make many observations and conduct investigations and experiments. You will also research information that is available from many sources. These activities will involve organizing and recording data. The quality of the data you collect and the way you organize it will determine how well others can understand and use it. In **Figure 1,** the student is obtaining and recording information using a thermometer.

Putting your observations in writing is an important way of communicating to others the information you have found and the results of your investigations and experiments.

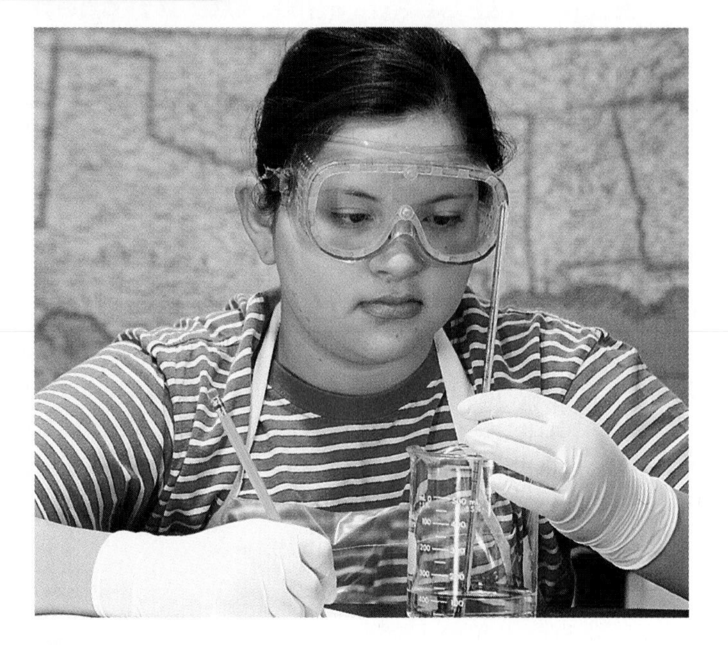

Figure 1
Making an observation is one way to gather information directly.

Researching Information

Scientists work to build on and add to human knowledge of the world. Before moving in a new direction, it is important to gather the information that already is known about a subject. You will look for such information in various reference sources. Follow these steps to research information on a scientific subject:

Step 1 Determine exactly what you need to know about the subject. For instance, you might want to find out about one of the elements in the periodic table.

Step 2 Make a list of questions, such as: Who discovered the element? When was it discovered? What makes the element useful or interesting?

Step 3 Use multiple sources such as textbooks, encyclopedias, government documents, professional journals, science magazines, and the Internet.

Step 4 List where you found the sources. Make sure the sources you use are reliable and the most current available.

Evaluating Print and Nonprint Sources

Not all sources of information are reliable. Evaluate the sources you use for information, and use only those you know to be dependable. For example, suppose you want to find ways to make your home more energy efficient. You might find two Web sites on how to save energy in your home. One Web site contains "Energy-Saving Tips" written by a company that sells a new type of weatherproofing material you put around your door frames. The other is a Web page on "Conserving Energy in Your Home" written by the U.S. Department of Energy. You would choose the second Web site as the more reliable source of information.

In science, information can change rapidly. Always consult the most current sources. A 1985 source about saving energy would not reflect the most recent research and findings.

Interpreting Scientific Illustrations

As you research a science topic, you will see drawings, diagrams, and photographs. Illustrations help you understand what you read. Some illustrations are included to help you understand an idea that you can't see easily by yourself. For instance, you can't see the tiny particles in an atom, but you can look at a diagram of an atom as labeled in **Figure 2** that helps you understand something about it. Visualizing a drawing helps many people remember details more easily. Illustrations also provide examples that clarify difficult concepts or give additional information about the topic you are studying.

Most illustrations have a label or caption. A label or caption identifies the illustration or provides additional information to better explain it. Can you find the caption or labels in **Figure 2?**

Figure 2
This drawing shows an atom of carbon with its six protons, six neutrons, and six electrons.

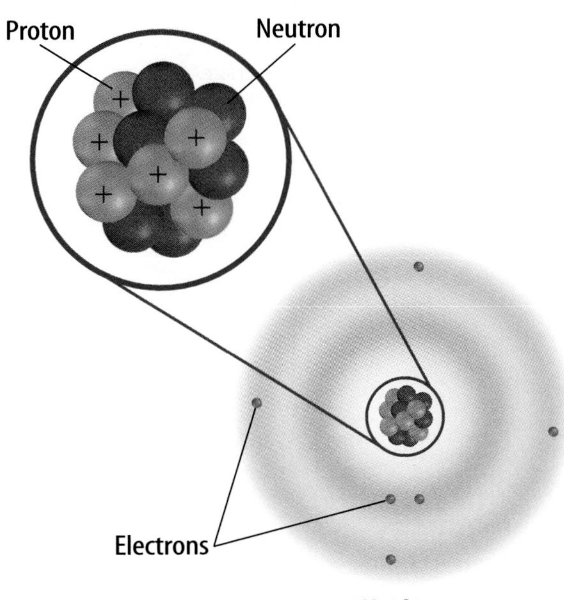

Concept Mapping

If you were taking a car trip, you might take some sort of road map. By using a map, you begin to learn where you are in relation to other places on the map.

A concept map is similar to a road map, but a concept map shows relationships among ideas (or concepts) rather than places. It is a diagram that visually shows how concepts are related. Because a concept map shows relationships among ideas, it can make the meanings of ideas and terms clear and help you understand what you are studying.

Overall, concept maps are useful for breaking large concepts down into smaller parts, making learning easier.

Venn Diagram

Although it is not a concept map, a Venn diagram illustrates how two subjects compare and contrast. In other words, you can see the characteristics that the subjects have in common and those that they do not.

The Venn diagram in **Figure 3** shows the relationship between two different substances made from the element carbon. However, due to the way their atoms are arranged, one substance is the gemstone diamond, and the other is the graphite found in pencils.

Figure 3
A Venn diagram shows how objects or concepts are alike and how they are different.

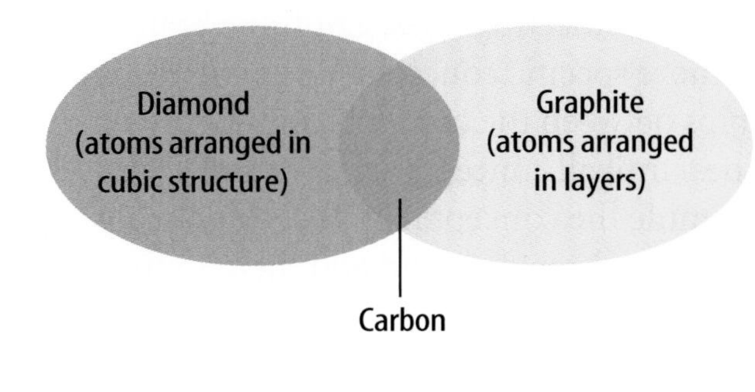

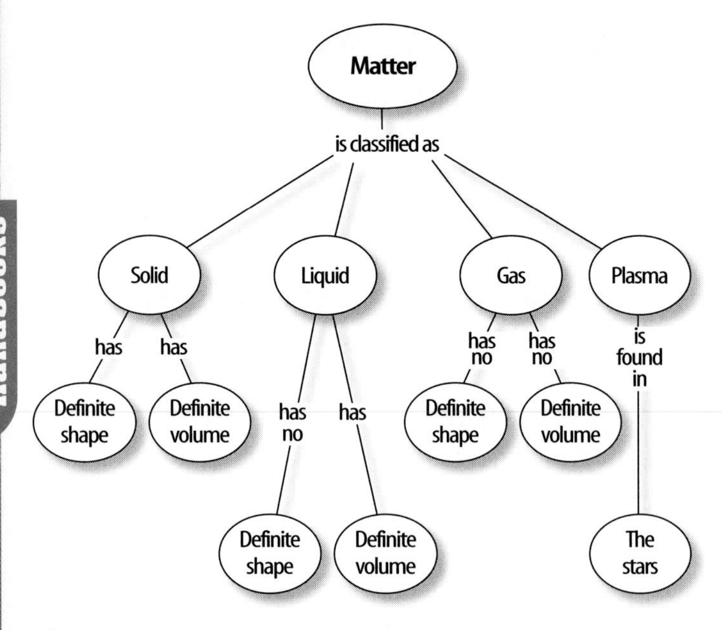

Figure 4
A network tree shows how concepts or objects are related.

Network Tree Look at the concept map in **Figure 4,** that describes the different types of matter. This is called a network tree concept map. Notice how some words are in ovals while others are written across connecting lines. The words inside the ovals are science terms or concepts. The words written on the connecting lines describe the relationships between the concepts.

When constructing a network tree, write the topic on a note card or piece of paper. Write the major concepts related to that topic on separate note cards or pieces of paper. Then arrange them in order from general to specific. Branch the related concepts from the major concept and describe the relationships on the connecting lines. Continue branching to more specific concepts. Write the relationships between the concepts on the connecting lines until all concepts are mapped. Then examine the concept map for relationships that cross branches, and add them to the concept map.

Events Chain An events chain is another type of concept map. It models the order of items or their sequence. In science, an events chain can be used to describe a sequence of events, the steps in a procedure, or the stages of a process.

When making an events chain, first find the one event that starts the chain. This event is called the *initiating event.* Then, find the next event in the chain and continue until you reach an outcome. Suppose you are asked to describe why and how a sound might make an echo. You might draw an events chain such as the one in **Figure 5.** Notice that connecting words are not necessary in an events chain.

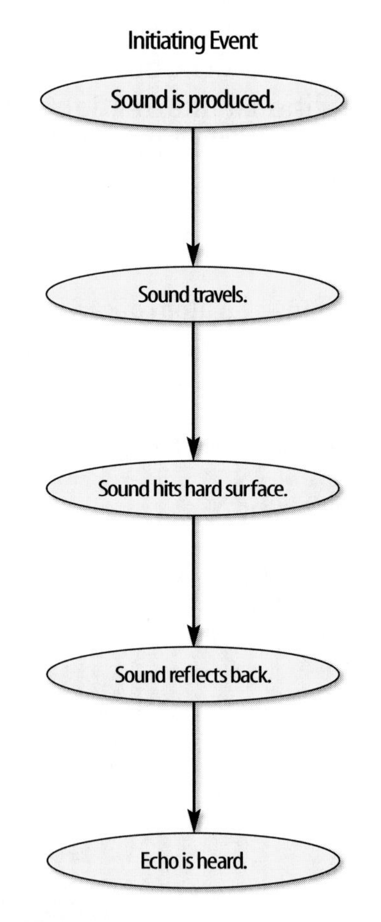

Figure 5
Events chains show the order of steps in a process or event.

Cycle Map A cycle concept map is a specific type of events chain map. In a cycle concept map, the series of events does not produce a final outcome. Instead, the last event in the chain relates back to the beginning event.

You first decide what event will be used as the beginning event. Once that is decided, you list events in order that occur after it. Words are written between events that describe what happens from one event to the next. The last event in a cycle concept map relates back to the beginning event. The number of events in a cycle concept varies, but is usually three or more. Look at the cycle map, as shown in **Figure 6.**

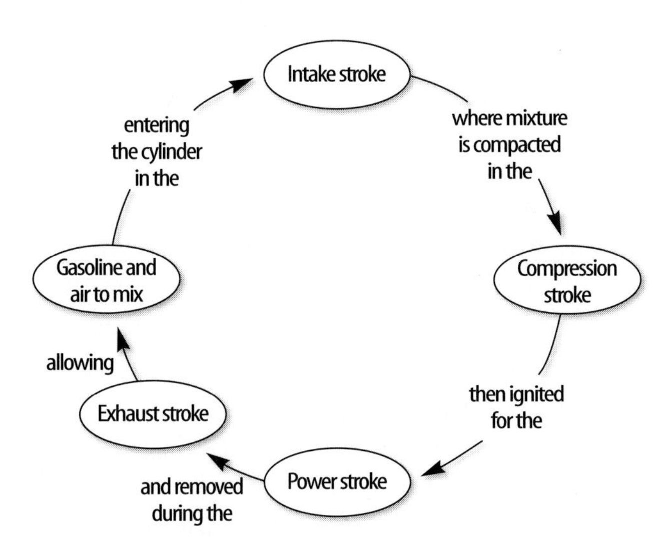

Figure 6
A cycle map shows events that occur in a cycle.

Spider Map A type of concept map that you can use for brainstorming is the spider map. When you have a central idea, you might find you have a jumble of ideas that relate to it but are not necessarily clearly related to each other. The spider map on sound in **Figure 7** shows that if you write these ideas outside the main concept, then you can begin to separate and group unrelated terms so they become more useful.

Figure 7
A spider map allows you to list ideas that relate to a central topic but not necessarily to one another.

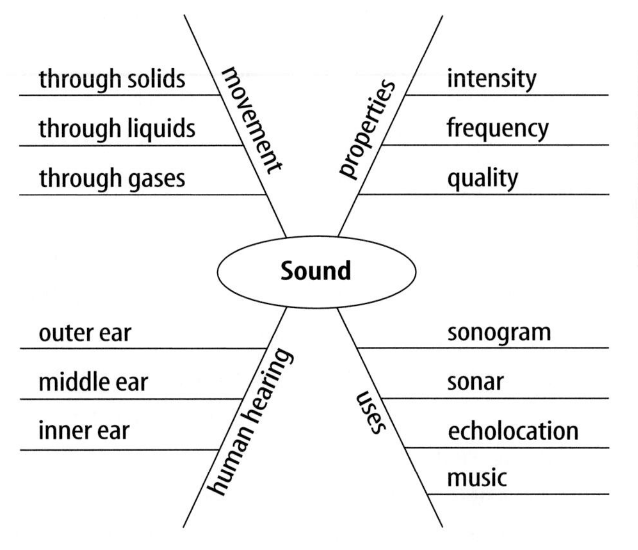

Skill Handbooks

Writing a Paper

You will write papers often when researching science topics or reporting the results of investigations or experiments. Scientists frequently write papers to share their data and conclusions with other scientists and the public. When writing a paper, use these steps.

Step 1 Assemble your data by using graphs, tables, or a concept map. Create an outline.
Step 2 Start with an introduction that contains a clear statement of purpose and what you intend to discuss or prove.
Step 3 Organize the body into paragraphs. Each paragraph should start with a topic sentence, and the remaining sentences in that paragraph should support your point.
Step 4 Position data to help support your points.
Step 5 Summarize the main points and finish with a conclusion statement.
Step 6 Use tables, graphs, charts, and illustrations whenever possible.

Investigating and Experimenting

You might say the work of a scientist is to solve problems. When you decide to find out why your neighbor's hydrangeas produce blue flowers while yours are pink, you are problem solving, too. You might also observe that your neighbor's azaleas are healthier than yours are and decide to see whether differences in the soil explain the differences in these plants.

Scientists use orderly approaches to solve problems. The methods scientists use include identifying a question, making observations, forming a hypothesis, testing a hypothesis, analyzing results, and drawing conclusions.

Scientific investigations involve careful observation under controlled conditions. Such observation of an object or a process can suggest new and interesting questions about it. These questions sometimes lead to the formation of a hypothesis. Scientific investigations are designed to test a hypothesis.

Identifying a Question

The first step in a scientific investigation or experiment is to identify a question to be answered or a problem to be solved. You might be interested in knowing how beams of laser light like the ones in **Figure 8** look the way they do.

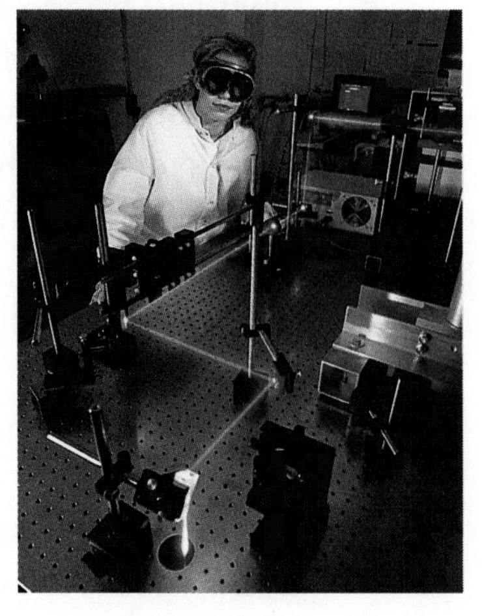

Figure 8
When you see lasers being used for scientific research, you might ask yourself, "Are these lasers different from those that are used for surgery?"

Forming Hypotheses

Hypotheses are based on observations that have been made. A hypothesis is a possible explanation based on previous knowledge and observations.

Perhaps a scientist has observed that certain substances dissolve faster in warm water than in cold. Based on these observations, the scientist can make a statement that he or she can test. The statement is a hypothesis. The hypothesis could be: *A substance dissolves in warm water faster.* A hypothesis has to be something you can test by using an investigation. A testable hypothesis is a valid hypothesis.

Predicting

When you apply a hypothesis, or general explanation, to a specific situation, you predict something about that situation. First, you must identify which hypothesis fits the situation you are considering. People use predictions to make everyday decisions. Based on previous observations and experiences, you might form a prediction that if substances dissolve in warm water faster, then heating the water will shorten mixing time for powdered fruit drinks. Someone could use this prediction to save time in preparing a fruit punch for a party.

Testing a Hypothesis

To test a hypothesis, you need a procedure. A procedure is the plan you follow in your experiment. A procedure tells you what materials to use, as well as how and in what order to use them. When you follow a procedure, data are generated that support or do not support the original hypothesis statement.

For example, premium gasoline costs more than regular gasoline. Does premium gasoline increase the efficiency or fuel mileage of your family car? You decide to test the hypothesis: "If premium gasoline is more efficient, then it should increase the fuel mileage of my family's car." Then you write the procedure shown in **Figure 9** for your experiment and generate the data presented in the table below.

Figure 9
A procedure tells you what to do step by step.

Procedure

1. Use regular gasoline for two weeks.
2. Record the number of kilometers between fill-ups and the amount of gasoline used.
3. Switch to premium gasoline for two weeks.
4. Record the number of kilometers between fill-ups and the amount of gasoline used.

Gasoline Data

Type of Gasoline	Kilometers Traveled	Liters Used	Liters per Kilometer
Regular	762	45.34	0.059
Premium	661	42.30	0.064

These data show that premium gasoline is less efficient than regular gasoline in one particular car. It took more gasoline to travel 1 km (0.064) using premium gasoline than it did to travel 1 km using regular gasoline (0.059). This conclusion does not support the hypothesis.

Are all investigations alike? Keep in mind as you perform investigations in science that a hypothesis can be tested in many ways. Not every investigation makes use of all the ways that are described on these pages, and not all hypotheses are tested by investigations. Scientists encounter many variations in the methods that are used when they perform experiments. The skills in this handbook are here for you to use and practice.

Identifying and Manipulating Variables and Controls

In any experiment, it is important to keep everything the same except for the item you are testing. The one factor you change is called the independent variable. The factor that changes as a result of the independent variable is called the dependent variable. Always make sure you have only one independent variable. If you allow more than one, you will not know what causes the changes you observe in the dependent variable. Many experiments also have controls—individual instances or experimental subjects for which the independent variable is not changed. You can then compare the test results to the control results.

For example, in the fuel-mileage experiment, you made everything the same except the type of gasoline that was used. The driver, the type of automobile, and the type of driving were the same throughout. In this way, you could be sure that any mileage differences were caused by the type of fuel—the independent variable. The fuel mileage was the dependent variable.

If you could repeat the experiment using several automobiles of the same type on a standard driving track with the same driver, you could make one automobile a control by using regular gasoline over the four-week period.

Collecting Data

Whether you are carrying out an investigation or a short observational experiment, you will collect data, or information. Scientists collect data accurately as numbers and descriptions and organize it in specific ways.

Observing Scientists observe items and events, then record what they see. When they use only words to describe an observation, it is called qualitative data. For example, a scientist might describe the color, texture, or odor of a substance produced in a chemical reaction. Scientists' observations also can describe how much there is of something. These observations use numbers, as well as words, in the description and are called quantitative data. For example, if a sample of the element gold is described as being "shiny and very dense," the data are clearly qualitative. Quantitative data on this sample of gold might include "a mass of 30 g and a density of 19.3 g/cm^3." Quantitative data often are organized into tables. Then, from information in the table, a graph can be drawn. Graphs can reveal relationships that exist in experimental data.

When you make observations in science, you should examine the entire object or situation first, then look carefully for details. If you're looking at an element sample, for instance, check the general color and pattern of the sample before using a hand lens to examine its surface for any smaller details or characteristics. Remember to record accurately everything you see.

Scientists try to make careful and accurate observations. When possible, they use instruments such as microscopes, metric rulers, graduated cylinders, thermometers, and balances. Measurements provide numerical data that can be repeated and checked.

Sampling When working with large numbers of objects or a large population, scientists usually cannot observe or study every one of them. Instead, they use a sample or a portion of the total number. To *sample* is to take a small, representative portion of the objects or organisms of a population for research. By making careful observations or manipulating variables within a portion of a group, information is discovered and conclusions are drawn that might apply to the whole population.

Estimating Scientific work also involves estimating. To estimate is to make a judgment about the size or the number of something without measuring or counting every object or member of a population. Scientists first measure or count the amount or number in a small sample. A geologist, for example, might remove a 10-g sample from a large rock that is rich in copper ore, as in **Figure 10.** Then a chemist would determine the percentage of copper by mass and multiply that percentage by the total mass of the rock to estimate the total mass of copper in the large rock.

Figure 10
Determining the percentage of copper by mass that is present in a small piece of a large rock, which is rich in copper ore, can help estimate the total mass of copper ore that is present in the rock.

Measuring in SI

The metric system of measurement was developed in 1795. A modern form of the metric system, called the International System, or SI, was adopted in 1960. SI provides standard measurements that all scientists around the world can understand.

The metric system is convenient because unit sizes vary by multiples of 10. When changing from smaller units to larger units, divide by a multiple of 10. When changing from larger units to smaller, multiply by a multiple of 10. To convert millimeters to centimeters, divide the millimeters by 10. To convert 30 mm to centimeters, divide 30 by 10 (30 mm equal 3 cm).

Prefixes are used to name units. Look at the table below for some common metric prefixes and their meanings. Do you see how the prefix *kilo-* attached to the unit *gram* is *kilogram*, or 1,000 g?

Metric Prefixes

Prefix	Symbol	Meaning	
kilo-	k	1,000	thousand
hecto-	h	100	hundred
deka-	da	10	ten
deci-	d	0.1	tenth
centi-	c	0.01	hundredth
milli-	m	0.001	thousandth

Now look at the metric ruler shown in **Figure 11.** The centimeter lines are the long, numbered lines, and the shorter lines are millimeter lines.

When using a metric ruler, line up the 0-cm mark with the end of the object being measured, and read the number of the unit where the object ends, in this instance it would be 4.5 cm.

Figure 11
This metric ruler has centimeter and millimeter divisions.

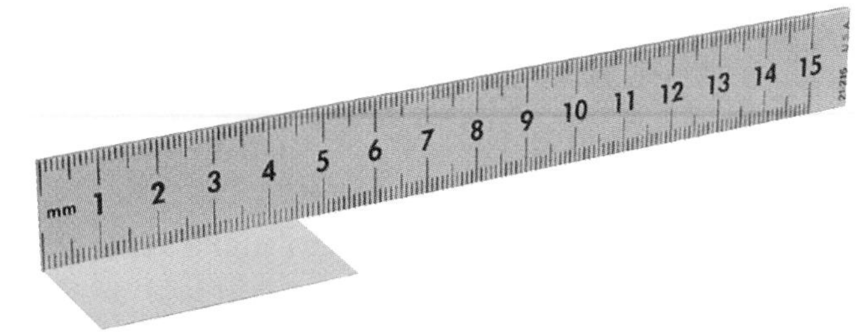

Liquid Volume In some science activities, you will measure liquids. The unit that is used to measure liquids is the liter. A liter has the volume of 1,000 cm^3. The prefix *milli-* means "thousandth (0.001)." A milliliter is one thousandth of 1 L, and 1 L has the volume of 1,000 mL. One milliliter of liquid completely fills a cube measuring 1 cm on each side. Therefore, 1 mL equals 1 cm^3.

You will use beakers and graduated cylinders to measure liquid volume. A graduated cylinder, as illustrated in **Figure 12,** is marked from bottom to top in milliliters. This one contains 79 mL of a liquid.

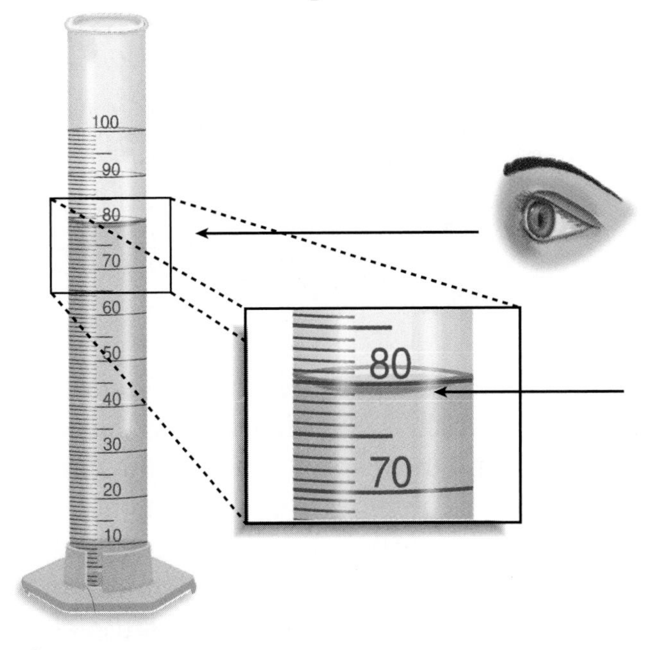

Figure 12
Graduated cylinders measure liquid volume.

Mass Scientists measure mass in grams. You might use a beam balance similar to the one shown in **Figure 13.** The balance has a pan on one side and a set of beams on the other side. Each beam has a rider that slides on the beam.

Before you find the mass of an object, slide all the riders back to the zero point. Check the pointer on the right to make sure it swings an equal distance above and below the zero point. If the swing is unequal, find and turn the adjusting screw until you have an equal swing.

Place an object on the pan. Slide the largest rider along its beam until the pointer drops below zero. Then move it back one notch. Repeat the process on each beam until the pointer swings an equal distance above and below the zero point. Sum the masses on each beam to find the mass of the object. Move all riders back to zero when finished.

Figure 13
A triple beam balance is used to determine the mass of an object.

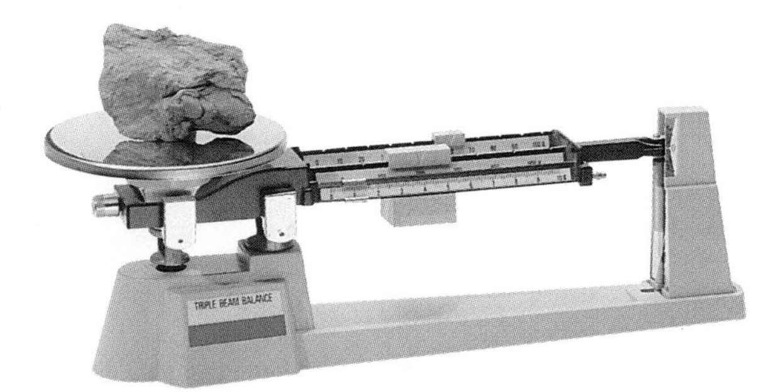

You should never place a hot object on the pan or pour chemicals directly onto the pan. Instead, find the mass of a clean container. Remove the container from the pan, then place the chemicals in the container. Find the mass of the container with the chemicals in it. To find the mass of the chemicals, subtract the mass of the empty container from the mass of the filled container.

Making and Using Tables

Browse through your textbook and you will see tables in the text and in the activities. In a table, data, or information, are arranged so that they are easier to understand. Activity tables help organize the data you collect during an activity so results can be interpreted.

Making Tables To make a table, list the items to be compared in the first column and the characteristics to be compared in the first row. The title should clearly indicate the content of the table, and the column or row heads should tell the reader what information is found in there. The table below lists materials collected for recycling on three weekly pick-up days. The inclusion of kilograms in parentheses also identifies for the reader that the figures are mass units.

Recyclable Materials Collected During Week

Day of Week	Paper (kg)	Aluminum (kg)	Glass (kg)
Monday	5.0	4.0	12.0
Wednesday	4.0	1.0	10.0
Friday	2.5	2.0	10.0

Using Tables How much paper, in kilograms, is being recycled on Wednesday? Locate the column labeled "Paper (kg)" and the row "Wednesday." The information in the box where the column and row intersect is the answer. Did you answer "4.0"? How much aluminum, in kilograms, is being recycled on Friday? If you answered "2.0," you understand how to read the table. How much glass is collected for recycling each week? Locate the column labeled "Glass (kg)" and add the figures for all three rows. If you answered "32.0," then you know how to locate and use the data provided in the table.

Recording Data

To be useful, the data you collect must be recorded carefully. Accuracy is key. A well-thought-out experiment includes a way to record procedures, observations, and results accurately. Data tables are one way to organize and record results. Set up the tables you will need ahead of time so you can record the data right away.

Record information properly and neatly. Never put unidentified data on scraps of paper. Instead, data should be written in a notebook like the one in **Figure 14.** Write in pencil so information isn't lost if your data gets wet. At each point in the experiment, record your data and label it. That way, your information will be accurate and you will not have to determine what the figures mean when you look at your notes later.

Figure 14
Record data neatly and clearly so it is easy to understand.

Recording Observations

It is important to record observations accurately and completely. That is why you always should record observations in your notes immediately as you make them. It is easy to miss details or make mistakes when recording results from memory. Do not include your personal thoughts when you record your data. Record only what you observe to eliminate bias. For example, when you record the time required for five students to climb the same set of stairs, you would note which student took the longest time. However, you would not refer to that student's time as "the worst time of all the students in the group."

Making Models

You can organize the observations and other data you collect and record in many ways. Making models is one way to help you better understand the parts of a structure you have been observing or the way a process for which you have been taking various measurements works.

Models often show things that are too large or too small for normal viewing. For example, you normally won't see the inside of an atom. However, you can understand the structure of the atom better by making a three-dimensional model of an atom. The relative sizes, the positions, and the movements of protons, neutrons, and electrons can be explained in words. An atomic model made of a plastic-ball nucleus and pipe-cleaner electron shells can help you visualize how the parts of the atom relate to each other.

Other models can be devised on a computer. Some models, such as those that illustrate the chemical combinations of different elements, are mathematical and are represented by equations.

Making and Using Graphs

After scientists organize data in tables, they might display the data in a graph that shows the relationship of one variable to another. A graph makes interpretation and analysis of data easier. Three types of graphs are the line graph, the bar graph, and the circle graph.

Line Graphs A line graph like in **Figure 15** is used to show the relationship between two variables. The variables being compared go on two axes of the graph. For data from an experiment, the independent variable always goes on the horizontal axis, called the *x*-axis. The dependent variable always goes on the vertical axis, called the *y*-axis. After drawing your axes, label each with a scale. Next, plot the data points.

A data point is the intersection of the recorded value of the dependent variable for each tested value of the independent variable. After all the points are plotted, connect them.

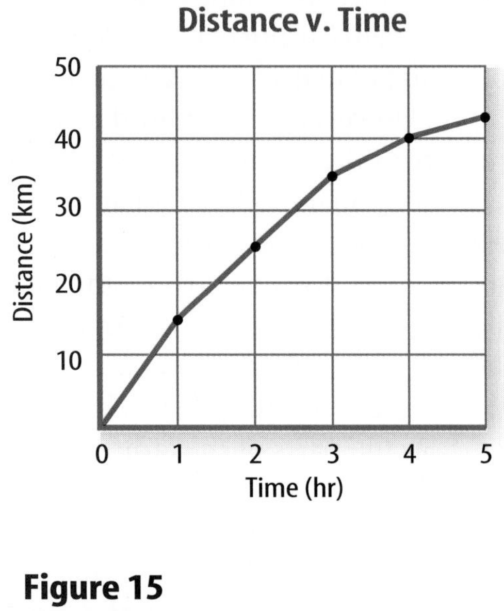

Distance v. Time
Distance (km)
Time (hr)

Figure 15
This line graph shows the relationship between distance and time during a bicycle ride lasting several hours.

Bar Graphs Bar graphs compare data that do not change continuously. Vertical bars show the relationships among data.

To make a bar graph, set up the *y*-axis as you did for the line graph. Draw vertical bars of equal size from the *x*-axis up to the point on the *y*-axis that represents value of *x*.

Figure 16
The amount of aluminum collected for recycling during one week can be shown as a bar graph or circle graph.

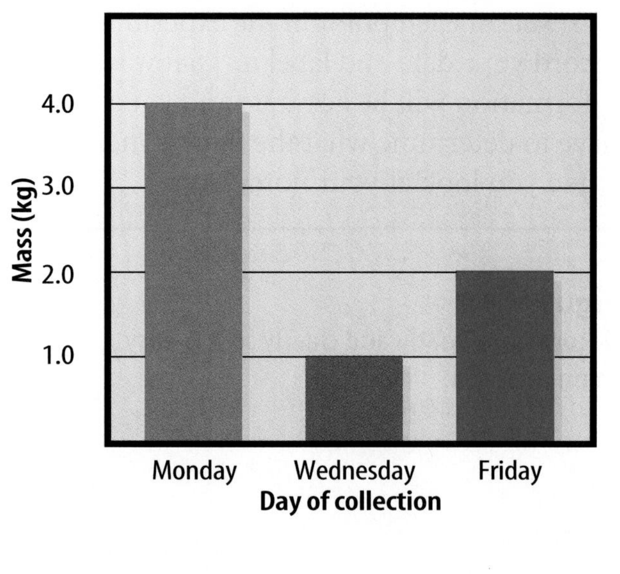

Circle Graphs A circle graph uses a circle divided into sections to display data as parts (fractions or percentages) of a whole. The size of each section corresponds to the fraction or percentage of the data that the section represents. So, the entire circle represents 100 percent, one-half represents 50 percent, one-fifth represents 20 percent, and so on.

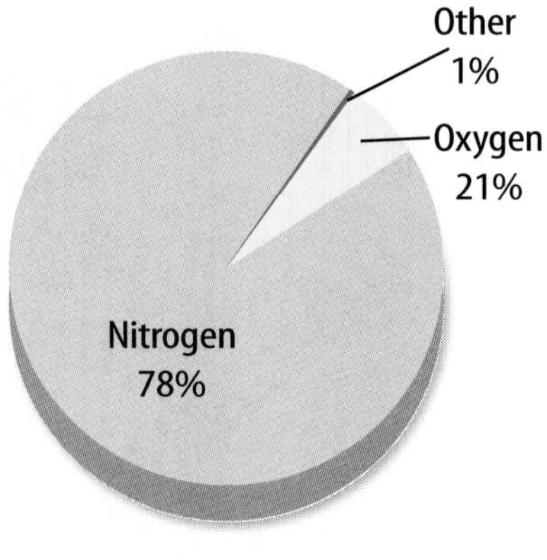

Other 1%
Oxygen 21%
Nitrogen 78%

Analyzing and Applying Results

Analyzing Results

To determine the meaning of your observations and investigation results, you will need to look for patterns in the data. You can organize your information in several of the ways that are discussed in this handbook. Then you must think critically to determine what the data mean. Scientists use several approaches when they analyze the data they have collected and recorded. Each approach is useful for identifying specific patterns in the data.

Forming Operational Definitions

An operational definition defines an object by showing how it functions, works, or behaves. Such definitions are written in terms of how an object works or how it can be used; that is, they describe its job or purpose.

For example, a ruler can be defined as a tool that measures the length of an object (how it can be used). A ruler also can be defined as something that contains a series of marks that can be used as a standard when measuring (how it works).

Classifying

Classifying is the process of sorting objects or events into groups based on common features. When classifying, first observe the objects or events to be classified. Then select one feature that is shared by some members in the group but not by all. Place those members that share that feature into a subgroup. You can classify members into smaller and smaller subgroups based on characteristics.

How might you classify a group of chemicals? You might first classify them by state of matter, putting solids, liquids, and gases into separate groups. Within each group, you could then look for another common feature by which to further classify members of the group, such as color or how reactive they are.

Remember that when you classify, you are grouping objects or events for a purpose. For example, classifying chemicals can be the first step in organizing them for storage. Both at home and at school, poisonous or highly reactive chemicals should all be stored in a safe location where they are not easily accessible to small children or animals. Solids, liquids, and gases each have specific storage requirements that may include waterproof, airtight, or pressurized containers. Are the dangerous chemicals in your home stored in the right place? Keep your purpose in mind as you select the features to form groups and subgroups.

Figure 17
Color is one of many characteristics that are used to classify chemicals.

Comparing and Contrasting

Observations can be analyzed by noting the similarities and differences between two or more objects or events that you observe. When you look at objects or events to see how they are similar, you are comparing them. Contrasting is looking for differences in objects or events. The table below compares and contrasts the characteristics of two elements.

Elemental Characteristics		
Element	Aluminum	Gold
Color	silver	gold
Classification	metal	metal
Density (g/cm^3)	2.7	19.3
Melting Point (°C)	660	1064

Recognizing Cause and Effect

Have you ever heard a loud pop right before the power went out and then suggested that an electric transformer probably blew out? If so, you have observed an effect and inferred a cause. The event is the effect, and the reason for the event is the cause.

When scientists are unsure of the cause of a certain event, they design controlled experiments to determine what caused it.

Interpreting Data

The word *interpret* means "to explain the meaning of something." Look at the problem originally being explored in an experiment and figure out what the data show. Identify the control group and the test group so you can see whether or not changes in the independent variable have had an effect. Look for differences in the dependent variable between the control and test groups.

These differences you observe can be qualitative or quantitative. You would be able to describe a qualitative difference using only words, whereas you would measure a quantitative difference and describe it using numbers. If there are differences, the independent variable that is being tested could have had an effect. If no differences are found between the control and test groups, the variable that is being tested apparently had no effect.

For example, suppose that three beakers each contain 100 mL of water. The beakers are placed on hot plates, and two of the hot plates are turned on, but the third is left off for a period of 5 min. Suppose you are then asked to describe any differences in the water in the three beakers. A qualitative difference might be the appearance of bubbles rising to the top in the water that is being heated but no rising bubbles in the unheated water. A quantitative difference might be a difference in the amount of water that is present in the beakers.

Inferring Scientists often make inferences based on their observations. An inference is an attempt to explain, or interpret, observations or to indicate what caused what you observed. An inference is a type of conclusion.

When making an inference, be certain to use accurate data and accurately described observations. Analyze all of the data that you've collected. Then, based on everything you know, explain or interpret what you've observed.

Drawing Conclusions

When scientists have analyzed the data they collected, they proceed to draw conclusions about what the data mean. These conclusions are sometimes stated using words similar to those found in the hypothesis formed earlier in the process.

Conclusions To analyze your data, you must review all of the observations and measurements that you made and recorded. Recheck all data for accuracy. After your data are rechecked and organized, you are almost ready to draw a conclusion such as "salt water boils at a higher temperature than freshwater."

Before you can draw a conclusion, however, you must determine whether the data allow you to come to a conclusion that supports a hypothesis. Sometimes that will be the case, other times it will not.

If your data do not support a hypothesis, it does not mean that the hypothesis is wrong. It means only that the results of the investigation did not support the hypothesis. Maybe the experiment needs to be redesigned, but very likely, some of the initial observations on which the hypothesis was based were incomplete or biased. Perhaps more observation or research is needed to refine the hypothesis.

Avoiding Bias Sometimes drawing a conclusion involves making judgments. When you make a judgment, you form an opinion about what your data mean. It is important to be honest and to avoid reaching a conclusion if there were no supporting evidence for it or if it were based on a small sample. It also is important not to allow any expectations of results to bias your judgments. If possible, it is a good idea to collect additional data. Scientists do this all the time.

For example, the *Hubble Space Telescope* was sent into space in April, 1990, to provide scientists with clearer views of the universe. The *Hubble* is the size of a school bus and has a 2.4-m-diameter mirror. The *Hubble* helped scientists answer questions about the planet Pluto.

For many years, scientists had only been able to hypothesize about the surface of the planet Pluto. The *Hubble* has now provided pictures of Pluto's surface that show a rough texture with light and dark regions on it. This might be the best information about Pluto scientists will have until they are able to send a space probe to it.

Evaluating Others' Data and Conclusions

Sometimes scientists have to use data that they did not collect themselves, or they have to rely on observations and conclusions drawn by other researchers. In cases such as these, the data must be evaluated carefully.

How were the data obtained? How was the investigation done? Was it carried out properly? Has it been duplicated by other researchers? Were they able to follow the exact procedure? Did they come up with the same results? Look at the conclusion, as well. Would you reach the same conclusion from these results? Only when you have confidence in the data of others can you believe it is true and feel comfortable using it.

Communicating

The communication of ideas is an important part of the work of scientists. A discovery that is not reported will not advance the scientific community's understanding or knowledge. Communication among scientists also is important as a way of improving their investigations.

Scientists communicate in many ways, from writing articles in journals and magazines that explain their investigations and experiments, to announcing important discoveries on television and radio, to sharing ideas with colleagues on the Internet or presenting them as lectures.

Computer Skills

People who study science rely on computers to record and store data and to analyze results from investigations. Whether you work in a laboratory or just need to write a lab report with tables, good computer skills are a necessity.

Using a Word Processor

Suppose your teacher has assigned a written report. After you've completed your research and decided how you want to write the information, you need to put all that information on paper. The easiest way to do this is with a word processing application on a computer.

A computer application that allows you to type your information, change it as many times as you need to, and then print it out so that it looks neat and clean is called a word processing application. You also can use this type of application to create tables and columns, add bullets or cartoon art to your page, include page numbers, and check your spelling.

Helpful Hints

- If you aren't sure how to do something using your word processing program, look in the help menu. You will find a list of topics there to click on for help. After you locate the help topic you need, just follow the step-by-step instructions you see on your screen.
- Just because you've spell checked your report doesn't mean that the spelling is perfect. The spell check feature can't catch misspelled words that look like other words. If you've accidentally typed *cold* instead of *gold*, the spell checker won't know the difference. Always reread your report to make sure you didn't miss any mistakes.

Figure 18
You can use computer programs to make graphs and tables.

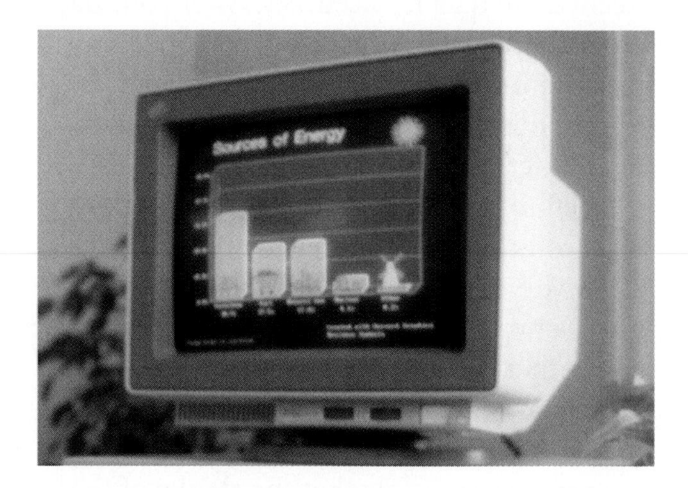

Using a Database

Imagine you're in the middle of a research project, busily gathering facts and information. You soon realize that it's becoming more difficult to organize and keep track of all the information. The tool to use to solve information overload is a database. Just as a file cabinet organizes paper records, a database organizes computer records. However, a database is more powerful than a simple file cabinet because at the click of a mouse, the contents can be reshuffled and reorganized. At computer-quick speeds, databases can sort information by any characteristics and filter data into multiple categories.

Helpful Hints

- Before setting up a database, take some time to learn the features of your database software by practicing with established database software.
- Periodically save your database as you enter data. That way, if something happens such as your computer malfunctions or the power goes off, you won't lose all of your work.

Doing a Database Search

When searching for information in a database, use the following search strategies to get the best results. These are the same search methods used for searching internet databases.

- Place the word *and* between two words in your search if you want the database to look for any entries that have both the words. For example, "gold *and* silver" would give you information that mentions both gold and silver.
- Place the word *or* between two words if you want the database to show entries that have at least one of the words. For example "gold *or* silver" would show you information that mentions either gold or silver.
- Place the word *not* between two words if you want the database to look for entries that have the first word but do not have the second word. For example, "gold *not* jewelry" would show you information that mentions gold but does not mention jewelry.

In summary, databases can be used to store large amounts of information about a particular subject. Databases allow biologists, Earth scientists, and physical scientists to search for information quickly and accurately.

Using an Electronic Spreadsheet

Your science fair experiment has produced lots of numbers. How do you keep track of all the data, and how can you easily work out all the calculations needed? You can use a computer program called a spreadsheet to record data that involve numbers. A spreadsheet is an electronic mathematical worksheet.

Type your data in rows and columns, just as they would look in a data table on a sheet of paper. A spreadsheet uses simple math to do data calculations. For example, you could add, subtract, divide, or multiply any of the values in the spreadsheet by another number. You also could set up a series of math steps you want to apply to the data. If you want to add 12 to all the numbers and then multiply all the numbers by 10, the computer does all the calculations for you in the spreadsheet. Below is an example of a spreadsheet that records test car data.

Helpful Hints

- Before you set up the spreadsheet, identify how you want to organize the data. Include any formulas you will need to use.
- Make sure you have entered the correct data into the correct rows and columns.
- You also can display your results in a graph. Pick the style of graph that best represents the data with which you are working.

Figure 19
A spreadsheet allows you to display large amounts of data and do calculations automatically.

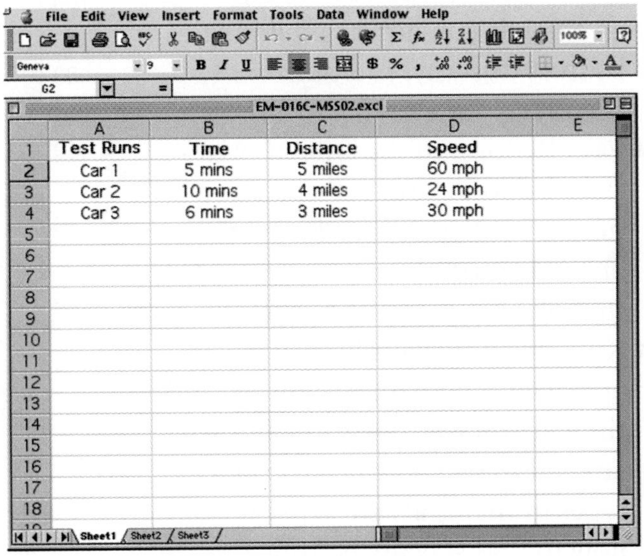

File Edit View Insert Format Tools Data Window Help

EM-016C-MSS02.excl

	A	B	C	D	E
1	Test Runs	Time	Distance	Speed	
2	Car 1	5 mins	5 miles	60 mph	
3	Car 2	10 mins	4 miles	24 mph	
4	Car 3	6 mins	3 miles	30 mph	
5					
6					
7					
8					
9					
10					
11					
12					
13					
14					
15					
16					
17					
18					

Sheet1 Sheet2 Sheet3

Using a Computerized Card Catalog

When you have a report or paper to research, you probably go to the library. To find the information you need in the library, you might have to use a computerized card catalog. This type of card catalog allows you to search for information by subject, by title, or by author. The computer then will display all the holdings the library has on the subject, title, or author requested.

A library's holdings can include books, magazines, databases, videos, and audio materials. When you have chosen something from this list, the computer will show whether an item is available and where in the library to find it.

Helpful Hints

- Remember that you can use the computer to search by subject, author, or title. If you know a book's author but not the title, you can search for all the books the library has by that author.
- When searching by subject, it's often most helpful to narrow your search by using specific search terms, such as *and, or,* and *not.* If you don't find enough sources this way, you can broaden your search.
- Pay attention to the type of materials found in your search. If you need a book, you can eliminate any videos or other resources that come up in your search.
- Knowing how your library is arranged can save you a lot of time. If you need help, the librarian will show you where certain types of materials are kept and how to find specific holdings.

Using Graphics Software

Are you having trouble finding that exact piece of art you're looking for? Do you have a picture in your mind of what you want but can't seem to find the right graphic to represent your ideas? To solve these problems, you can use graphics software. Graphics software allows you to create and change images and diagrams in almost unlimited ways. Typical uses for graphics software include arranging clip art, changing scanned images, and constructing pictures from scratch. Most graphics software applications work in similar ways. They use the same basic tools and functions. Once you master one graphics application, you can use other graphics applications.

Figure 20
Graphics software can use your data to draw bar graphs.

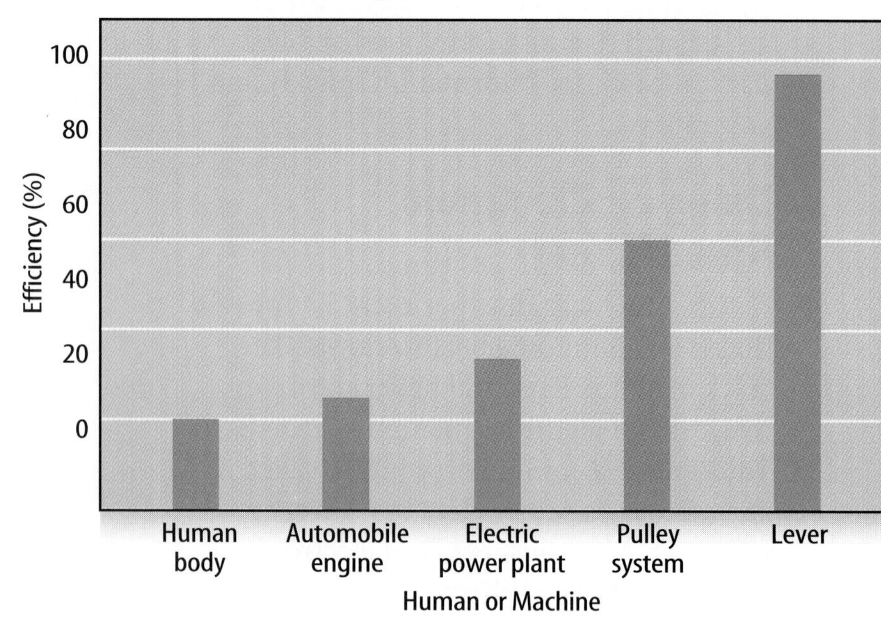

Figure 21
Graphics software can use your data to draw circle graphs.

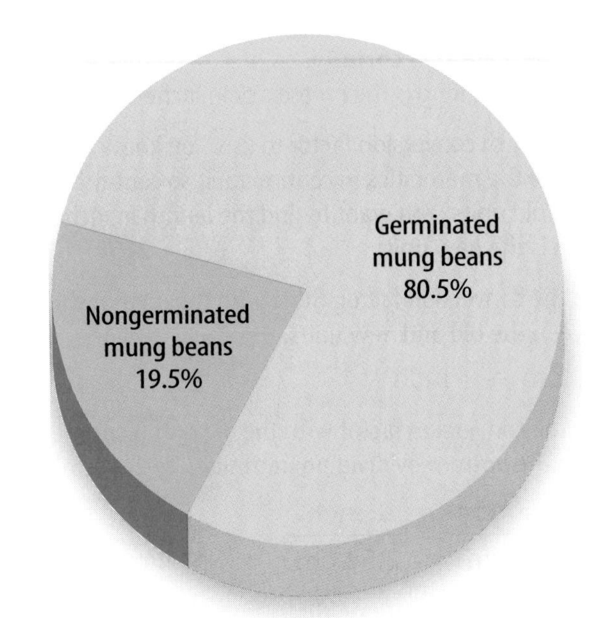

Helpful Hints

- As with any method of drawing, the more you practice using the graphics software, the better your results will be.
- Start by using the software to manipulate existing drawings. Once you master this, making your own illustrations will be easier.
- Clip art is available on CD-ROMs and the Internet. With these resources, finding a piece of clip art to suit your purposes is simple.
- As you work on a drawing, save it often.

Developing Multimedia Presentations

It's your turn—you have to present your science report to the entire class. How do you do it? You can use many different sources of information to get the class excited about your presentation. Posters, videos, photographs, sound, computers, and the Internet can help show your ideas.

First, determine what important points you want to make in your presentation. Then, write an outline of what materials and types of media would best illustrate those points. Maybe you could start with an outline on an overhead projector, then show a video, followed by something from the Internet or a slide show accompanied by music or recorded voices. You might choose to use a presentation builder computer application that can combine all these elements into one presentation. Make sure the presentation is well constructed to make the most impact on the audience.

Figure 22
Multimedia presentations use many types of print and electronic materials.

Helpful Hints

- Carefully consider what media will best communicate the point you are trying to make.
- Make sure you know how to use any equipment you will be using in your presentation.
- Practice the presentation several times.
- If possible, set up all of the equipment ahead of time. Make sure everything is working correctly.

Math Skill Handbook

Use this Math Skill Handbook to help solve problems you are given in this text. You might find it useful to review topics in this Math Skill Handbook first.

Skill Handbooks

Converting Units

In science, quantities such as length, mass, and time sometimes are measured using different units. Suppose you want to know how many miles are in 12.7 km?

Conversion factors are used to change from one unit of measure to another. A conversion factor is a ratio that is equal to one. For example, there are 1,000 mL in 1 L, so 1,000 mL equals 1 L, or:

$$1{,}000 \text{ mL} = 1 \text{ L}$$

If both sides are divided by 1 L, this equation becomes:

$$\frac{1{,}000 \text{ mL}}{1 \text{ L}} = 1$$

The **ratio** on the left side of this equation is equal to one and is a conversion factor. You can make another conversion factor by dividing both sides of the top equation by 1,000 mL:

$$1 = \frac{1 \text{ L}}{1{,}000 \text{ mL}}$$

To **convert units,** you multiply by the appropriate conversion factor. For example, how many milliliters are in 1.255 L? To convert 1.255 L to milliliters, multiply 1.255 L by a conversion factor.

Use the **conversion factor** with new units (mL) in the numerator and the old units (L) in the denominator.

$$1.255 \cancel{\text{L}} \times \frac{1{,}000 \text{ mL}}{1 \cancel{\text{L}}} = 1{,}255 \text{ mL}$$

The unit L divides in this equation, just as if it were a number.

Example 1 There are 2.54 cm in 1 inch. If a meterstick has a length of 100 cm, how long is the meterstick in inches?

Step 1 Decide which conversion factor to use. You know the length of the meterstick in centimeters, so centimeters are the old units. You want to find the length in inches, so inch is the new unit.

Step 2 Form the conversion factor. Start with the relationship between the old and new units.

$$2.54 \text{ cm} = 1 \text{ inch}$$

Step 3 Form the conversion factor with the old unit (centimeter) on the bottom by dividing both sides by 2.54 cm.

$$1 = \frac{2.54 \text{ cm}}{2.54 \text{ cm}} = \frac{1 \text{ inch}}{2.54 \text{ cm}}$$

Step 4 Multiply the old measurement by the conversion factor.

$$100 \cancel{\text{cm}} \times \frac{1 \text{ inch}}{2.54 \cancel{\text{cm}}} = 39.37 \text{ inches}$$

The meter stick is 39.37 inches long.

Example 2 There are 365 days in one year. If a person is 14 years old, what is his or her age in days? (Ignore leap years)

Step 1 Decide which conversion factor to use. You want to convert years to days.

Step 2 Form the conversion factor. Start with the relation between the old and new units.

$$1 \text{ year} = 365 \text{ days}$$

Step 3 Form the conversion factor with the old unit (year) on the bottom by dividing both sides by 1 year.

$$1 = \frac{1 \text{ year}}{1 \text{ year}} = \frac{365 \text{ days}}{1 \text{ year}}$$

Step 4 Multiply the old measurement by the conversion factor:

$$14 \cancel{\text{years}} \times \frac{365 \text{ days}}{1 \cancel{\text{year}}} = 5{,}110 \text{ days}$$

The person's age is 5,110 days.

Practice Problem A book has a mass of 2.31 kg. If there are 1,000 g in 1 kg, what is the mass of the book in grams?

Using Fractions

A **fraction** is a number that compares a part to the whole. For example, in the fraction $\frac{2}{3}$, the 2 represents the part and the 3 represents the whole. In the fraction $\frac{2}{3}$, the top number, 2, is called the numerator. The bottom number, 3, is called the denominator.

Sometimes fractions are not written in their simplest form. To determine a fraction's **simplest form,** you must find the greatest common factor (GCF) of the numerator and denominator. The greatest common factor is the largest factor that is common to the numerator and denominator.

For example, because the number 3 divides into 12 and 30 evenly, it is a common factor of 12 and 30. However, because the number 6 is the largest number that evenly divides into 12 and 30, it is the **greatest common factor.**

After you find the greatest common factor, you can write a fraction in its simplest form. Divide both the numerator and the denominator by the greatest common factor. The number that results is the fraction in its **simplest form.**

Example Twelve of the 20 chemicals used in the science lab are in powder form. What fraction of the chemicals used in the lab are in powder form?

Step 1 Write the fraction.

$$\frac{\text{part}}{\text{whole}} = \frac{12}{20}$$

Step 2 To find the GCF of the numerator and denominator, list all of the factors of each number.

Factors of 12: 1, 2, 3, 4, 6, 12 (the numbers that divide evenly into 12)

Factors of 20: 1, 2, 4, 5, 10, 20 (the numbers that divide evenly into 20)

Step 3 List the common factors.

1, 2, 4.

Step 4 Choose the greatest factor in the list of common factors.

The GCF of 12 and 20 is 4.

Step 5 Divide the numerator and denominator by the GCF.

$$\frac{12 \div 4}{20 \div 4} = \frac{3}{5}$$

In the lab, $\frac{3}{5}$ of the chemicals are in powder form.

Practice Problem There are 90 rides at an amusement park. Of those rides, 66 have a height restriction. What fraction of the rides has a height restriction? Write the fraction in simplest form.

Calculating Ratios

A **ratio** is a comparison of two numbers by division.

Ratios can be written 3 to 5 or 3:5. Ratios also can be written as fractions, such as $\frac{3}{5}$. Ratios, like fractions, can be written in simplest form. Recall that a fraction is in **simplest form** when the greatest common factor (GCF) of the numerator and denominator is 1.

Example A chemical solution contains 40 g of salt and 64 g of baking soda. What is the ratio of salt to baking soda as a fraction in simplest form?

Step 1 Write the ratio as a fraction. $\frac{\text{salt}}{\text{baking soda}} = \frac{40}{64}$

Step 2 Express the fraction in simplest form.
The GCF of 40 and 64 is 8.

$$\frac{40}{64} = \frac{40 \div 8}{64 \div 8} = \frac{5}{8}$$

The ratio of salt to baking soda in the solution is $\frac{5}{8}$.

Practice Problem Two metal rods measure 100 cm and 144 cm in length. What is the ratio of their lengths in simplest fraction form?

Using Decimals

A **decimal** is a fraction with a denominator of 10, 100, 1,000, or another power of 10. For example, 0.854 is the same as the fraction $\frac{854}{1,000}$.

In a decimal, the decimal point separates the ones place and the tenths place. For example, 0.27 means twenty-seven hundredths, or $\frac{27}{100}$, where 27 is the **number of units** out of 100 units. Any fraction can be written as a decimal using division.

Example Write $\frac{5}{8}$ as a decimal.

Step 1 Write a division problem with the numerator, 5, as the dividend and the denominator, 8, as the divisor. Write 5 as 5.000.

Step 2 Solve the problem.

$$\begin{array}{r} 0.625 \\ 8\overline{)5.000} \\ \underline{48} \\ 20 \\ \underline{16} \\ 40 \\ \underline{40} \\ 0 \end{array}$$

Therefore, $\frac{5}{8} = 0.625$.

Practice Problem Write $\frac{19}{25}$ as a decimal.

Using Percentages

The word *percent* means "out of one hundred." A **percent** is a ratio that compares a number to 100. Suppose you read that 77 percent of Earth's surface is covered by water. That is the same as reading that the fraction of Earth's surface covered by water is $\frac{77}{100}$. To express a fraction as a percent, first find an equivalent decimal for the fraction. Then, multiply the decimal by 100 and add the percent symbol. For example, $\frac{1}{2} = 1 \div 2 = 0.5$. Then $0.5 = 0.50 = 50\%$.

Example Express $\frac{13}{20}$ as a percent.

Step 1 Find the equivalent decimal for the fraction.

$$\begin{array}{r} 0.65 \\ 20\overline{)13.00} \\ \underline{120} \\ 100 \\ \underline{100} \\ 0 \end{array}$$

Step 2 Rewrite the fraction $\frac{13}{20}$ as 0.65.

Step 3 Multiply 0.65 by 100 and add the % sign.

$$0.65 \cdot 100 = 65 = 65\%$$

So, $\frac{13}{20} = 65\%$.

Practice Problem In one year, 73 of 365 days were rainy in one city. What percent of the days in that city were rainy?

Using Precision and Significant Digits

When you make a **measurement,** the value you record depends on the precision of the measuring instrument. When adding or subtracting numbers with different precision, the answer is rounded to the smallest number of decimal places of any number in the sum or difference. When multiplying or dividing, the answer is rounded to the smallest number of significant figures of any number being multiplied or divided. When counting the number of **significant figures,** all digits are counted except zeros at the end of a number with no decimal such as 2,500, and zeros at the beginning of a decimal such as 0.03020.

Example The lengths 5.28 and 5.2 are measured in meters. Find the sum of these lengths and report the sum using the least precise measurement.

Step 1 Find the sum.

$$\begin{array}{rl} 5.28 \text{ m} & \text{2 digits after the decimal} \\ +\ 5.2\ \ \text{m} & \text{1 digit after the decimal} \\ \hline 10.48 \text{ m} & \end{array}$$

Step 2 Round to one digit after the decimal because the least number of digits after the decimal of the numbers being added is 1.

The sum is 10.5 m.

Practice Problem Multiply the numbers in the example using the rule for multiplying and dividing. Report the answer with the correct number of significant figures.

Solving One-Step Equations

An **equation** is a statement that two things are equal. For example, $A = B$ is an equation that states that A is equal to B.

Sometimes one side of the equation will contain a **variable** whose value is not known. In the equation $3x = 12$, the variable is x.

The equation is solved when the variable is replaced with a value that makes both sides of the equation equal to each other. For example, the solution of the equation $3x = 12$ is $x = 4$. If the x is replaced with 4, then the equation becomes $3 \cdot 4 = 12$, or $12 = 12$.

To solve an equation such as $8x = 40$, divide both sides of the equation by the number that multiplies the variable.

$$8x = 40$$
$$\frac{8x}{8} = \frac{40}{8}$$
$$x = 5$$

You can check your answer by replacing the variable with your solution and seeing if both sides of the equation are the same.

$$8x = 8 \cdot 5 = 40$$

The left and right sides of the equation are the same, so $x = 5$ is the solution.

Sometimes an equation is written in this way: $a = bc$. This also is called a **formula.** The letters can be replaced by numbers, but the numbers must still make both sides of the equation the same.

Example 1 Solve the equation $10x = 35$.

Step 1 Find the solution by dividing each side of the equation by 10.

$$10x = 35 \qquad \frac{10x}{10} = \frac{35}{10} \qquad x = 3.5$$

Step 2 Check the solution.

$$10x = 35 \qquad 10 \times 3.5 = 35 \qquad 35 = 35$$

Both sides of the equation are equal, so $x = 3.5$ is the solution to the equation.

Example 2 In the formula $a = bc$, find the value of c if $a = 20$ and $b = 2$.

Step 1 Rearrange the formula so the unknown value is by itself on one side of the equation by dividing both sides by b.

$$a = bc$$
$$\frac{a}{b} = \frac{bc}{b}$$
$$\frac{a}{b} = c$$

Step 2 Replace the variables a and b with the values that are given.

$$\frac{a}{b} = c$$
$$\frac{20}{2} = c$$
$$10 = c$$

Step 3 Check the solution.

$$a = bc$$
$$20 = 2 \times 10$$
$$20 = 20$$

Both sides of the equation are equal, so $c = 10$ is the solution when $a = 20$ and $b = 2$.

Practice Problem In the formula $h = gd$, find the value of d if $g = 12.3$ and $h = 17.4$.

Using Proportions

A **proportion** is an equation that shows that two ratios are equivalent. The ratios $\frac{2}{4}$ and $\frac{5}{10}$ are equivalent, so they can be written as $\frac{2}{4} = \frac{5}{10}$. This equation is an example of a proportion.

When two ratios form a proportion, the **cross products** are equal. To find the cross products in the proportion $\frac{2}{4} = \frac{5}{10}$, multiply the 2 and the 10, and the 4 and the 5. Therefore **$2 \cdot 10 = 4 \cdot 5$, or $20 = 20$.**

Because you know that both proportions are equal, you can use cross products to find a missing term in a proportion. This is known as **solving the proportion.** Solving a proportion is similar to solving an equation.

Example The heights of a tree and a pole are proportional to the lengths of their shadows. The tree casts a shadow of 24 m at the same time that a 6-m pole casts a shadow of 4 m. What is the height of the tree?

Step 1 Write a proportion.

$$\frac{\text{height of tree}}{\text{height of pole}} = \frac{\text{length of tree's shadow}}{\text{length of pole's shadow}}$$

Step 2 Substitute the known values into the proportion. Let h represent the unknown value, the height of the tree.

$$\frac{h}{6} = \frac{24}{4}$$

Step 3 Find the cross products.

$$h \cdot 4 = 6 \cdot 24$$

Step 4 Simplify the equation.

$$4h = 144$$

Step 5 Divide each side by 4.

$$\frac{4h}{4} = \frac{144}{4}$$

$$h = 36$$

The height of the tree is 36 m.

Practice Problem The ratios of the weights of two objects on the Moon and on Earth are in proportion. A rock weighing 3 N on the Moon weighs 18 N on Earth. How much would a rock that weighs 5 N on the Moon weigh on Earth?

Using Statistics

Statistics is the branch of mathematics that deals with collecting, analyzing, and presenting data. In statistics, there are three common ways to summarize the data with a single number—the mean, the median, and the mode.

The **mean** of a set of data is the arithmetic average. It is found by adding the numbers in the data set and dividing by the number of items in the set.

The **median** is the middle number in a set of data when the data are arranged in numerical order. If there were an even number of data points, the median would be the mean of the two middle numbers.

The **mode** of a set of data is the number or item that appears most often.

Another number that often is used to describe a set of data is the range. The **range** is the difference between the largest number and the smallest number in a set of data.

A **frequency table** shows how many times each piece of data occurs, usually in a survey. The frequency table below shows the results of a student survey on favorite color.

Color	Tally	Frequency
red	||||	4
blue	~~||||~~	5
black	||	2
green	|||	3
purple	~~||||~~ ||	7
yellow	~~||||~~ |	6

Based on the frequency table data, which color is the favorite?

Example The speeds (in m/s) for a race car during five different time trials are 39, 37, 44, 36, and 44.

To find the mean:

Step 1 Find the sum of the numbers.

$$39 + 37 + 44 + 36 + 44 = 200$$

Step 2 Divide the sum by the number of items, which is 5.

$$200 \div 5 = 40$$

The mean measure is 40 m/s.

To find the median:

Step 1 Arrange the measures from least to greatest.

$$36,\ 37,\ \underline{39},\ 44,\ 44$$

Step 2 Determine the middle measure.

The median measure is 39 m/s.

To find the mode:

Step 1 Group the numbers that are the same together.

$$44, 44, 36, 37, 39$$

Step 2 Determine the number that occurs most in the set.

$$\underline{44, 44}, 36, 37, 39$$

The mode measure is 44 m/s.

To find the range:

Step 1 Arrange the measures from largest to smallest.

$$44, 44, 39, 37, 36$$

Step 2 Determine the largest and smallest measures in the set.

$$\underline{44}, 44, 39, 37, \underline{36}$$

Step 3 Find the difference between the largest and smallest measures.

$$44 - 36 = 8$$

The range is 8 m/s.

Practice Problem Find the mean, median, mode, and range for the data set 8, 4, 12, 8, 11, 14, 16.

Safety in the Science Classroom

1. Always obtain your teacher's permission to begin an investigation.
2. Study the procedure. If you have questions, ask your teacher. Be sure you understand any safety symbols shown on the page.
3. Use the safety equipment provided for you. Goggles and a safety apron should be worn during most investigations.
4. Always slant test tubes away from yourself and others when heating them or adding substances to them.
5. Never eat or drink in the lab, and never use lab glassware as food or drink containers. Never inhale chemicals. Do not taste any substances or draw any material into a tube with your mouth.
6. Report any spill, accident, or injury, no matter how small, immediately to your teacher, then follow his or her instructions.
7. Know the location and proper use of the fire extinguisher, safety shower, fire blanket, first aid kit, and fire alarm.
8. Keep all materials away from open flames. Tie back long hair and tie down loose clothing.
9. If your clothing should catch fire, smother it with the fire blanket, or get under a safety shower. NEVER RUN.
10. If a fire should occur, turn off the gas then leave the room according to established procedures.

Follow these procedures as you clean up your work area

1. Turn off the water and gas. Disconnect electrical devices.
2. Clean all pieces of equipment and return all materials to their proper places.
3. Dispose of chemicals and other materials as directed by your teacher. Place broken glass and solid substances in the proper containers. Make sure never to discard materials in the sink.
4. Clean your work area. Wash your hands thoroughly after working in the laboratory.

First Aid

Injury	Safe Response ALWAYS NOTIFY YOUR TEACHER IMMEDIATELY
Burns	Apply cold water.
Cuts and Bruises	Stop any bleeding by applying direct pressure. Cover cuts with a clean dressing. Apply ice packs or cold compresses to bruises.
Fainting	Leave the person lying down. Loosen any tight clothing and keep crowds away.
Foreign Matter in Eye	Flush with plenty of water. Use eyewash bottle or fountain.
Poisoning	Note the suspected poisoning agent.
Any Spills on Skin	Flush with large amounts of water or use safety shower.

Reference Handbook B

SI—Metric/English, English/Metric Conversions

	When you want to convert:	To:	Multiply by:
Length	inches	centimeters	2.54
	centimeters	inches	0.39
	yards	meters	0.91
	meters	yards	1.09
	miles	kilometers	1.61
	kilometers	miles	0.62
Mass and Weight*	ounces	grams	28.35
	grams	ounces	0.04
	pounds	kilograms	0.45
	kilograms	pounds	2.2
	tons (short)	tonnes (metric tons)	0.91
	tonnes (metric tons)	tons (short)	1.10
	pounds	newtons	4.45
	newtons	pounds	0.22
Volume	cubic inches	cubic centimeters	16.39
	cubic centimeters	cubic inches	0.06
	liters	quarts	1.06
	quarts	liters	0.95
	gallons	liters	3.78
Area	square inches	square centimeters	6.45
	square centimeters	square inches	0.16
	square yards	square meters	0.83
	square meters	square yards	1.19
	square miles	square kilometers	2.59
	square kilometers	square miles	0.39
	hectares	acres	2.47
	acres	hectares	0.40
Temperature	To convert °Celsius to °Fahrenheit		°C × 9/5 + 32
	To convert °Fahrenheit to °Celsius		5/9 (°F − 32)

*Weight is measured in standard Earth gravity.

English Glossary

This glossary defines each key term that appears in bold type in the text. It also shows the chapter, section, and page number where you can find the word used.

A

acid: substance that releases H^+ ions and produces hydronium ions when dissolved in water. (Chap. 3, Sec. 3, p. 78)

activation energy: minimum amount of energy needed to start a chemical reaction. (Chap. 2, Sec. 2, p. 47)

amino (uh ME noh) **group:** consists of one nitrogen atom covalently bonded to two hydrogen atoms; represented by the formula –NH2. (Chap. 4, Sec. 2, p. 105)

amino acids: building blocks of proteins; contain both an amino group and a carboxyl acid group replacing hydrogens on the same carbon atom. (Chap. 4, Sec. 2, p. 105)

aqueous (A kwee us): solution in which water is the solvent. (Chap. 3, Sec. 2, p. 70)

B

base: substance that accepts H^+ ions and produces hydroxide ions when dissolved in water. (Chap. 3, Sec. 3, p. 81)

C

carbohydrates: organic compounds containing only carbon, hydrogen, and oxygen; starches, cellulose, glycogen, sugars. (Chap. 4, Sec. 3, p. 110)

carboxyl (car BOK sul) **group:** consists of one carbon atom, two oxygen atoms, and one hydrogen atom; represented by the formula –COOH. (Chap. 4, Sec. 2, p. 105)

catalyst: substance that speeds up a chemical reaction but is not used up itself or permanently changed. (Chap. 4, Sec. 2, p. 51)

chemical bond: force that holds two atoms together. (Chap. 1, Sec. 1, p. 15)

chemical equation: shorthand form for writing what reactants are used and what products are formed in a chemical reaction; sometimes shows whether energy is produced or absorbed. (Chap. 2, Sec. 1, p. 38)

chemical formula: combination of chemical symbols and numbers that indicates which elements and how many atoms of each element are present in a molecule. (Chap. 1, Sec. 2, p. 24)

chemical reaction: process that produces chemical change, resulting in new substances that have properties different from those of the original substances. (Chap. 2, Sec. 1, p. 36)

compound: pure substance that contains two or more elements. (Chap. 1, Sec. 2, p. 17)

concentration: describes how much solute is present in a solution compared to the amount of solvent. (Chap. 3, Sec. 2, p. 75)

covalent bond: chemical bond formed when atoms share electrons. (Chap. 1, Sec. 2, p. 19)

E

electron cloud: area where negatively charged electrons, arranged in energy levels, travel around an atom's nucleus. (Chap. 1, Sec. 1, p. 8)

electron dot diagram: chemical symbol for an element; surrounded by as many dots as there are electrons in its outer energy level. (Chap. 1, Sec. 1, p. 14)

endothermic (en duh THUR mihk) **reaction:** chemical reaction in which heat energy is absorbed. (Chap. 2, Sec. 1, p. 43)

exothermic (ek soh THUR mihk) **reaction:** chemical reaction in which heat energy is released. (Chap. 2, Sec. 1, p. 43)

H

hydrocarbon: organic compound that has only carbon and hydrogen atoms. (Chap. 4, Sec. 1, p. 97)

hydroxyl (hi DROK sul) **group:** consists of an oxygen atom and a hydrogen atom joined by a covalent bond; represented by the formula –OH. (Chap. 4, Sec. 2, p. 104)

I

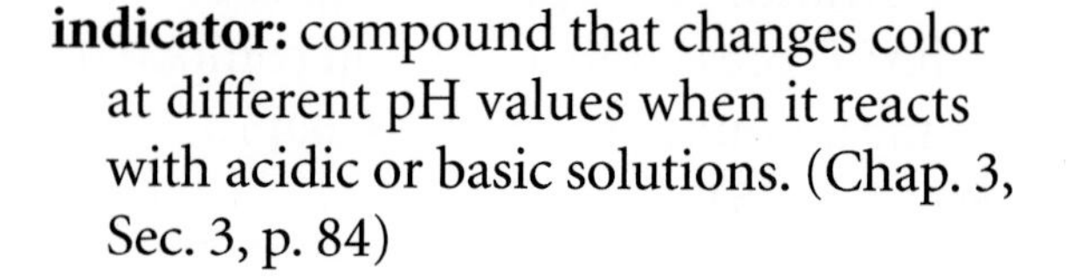

indicator: compound that changes color at different pH values when it reacts with acidic or basic solutions. (Chap. 3, Sec. 3, p. 84)

inhibitor: substance that slows down a chemical reaction, making the formation of a certain amount of product take longer. (Chap. 2, Sec. 2, p. 50)

ion (I ahn): atom that is no longer neutral because it has gained or lost an electron. (Chap. 1, Sec. 2, p. 17)

ionic bond: attraction that holds oppositely charged ions close together. (Chap. 1, Sec. 2, p. 17)

isomers (I suh murz): compounds with the same chemical formula but different structures and different physical and chemical properties. (Chap. 4, Sec. 1, p. 100)

L

lipids: organic compounds made up of three long-chain carboxylic acids bonded to glycerol; fats, oils, waxes. (Chap. 4, Sec. 3, p. 113)

M

molecule (MAH lih kewl): neutral particle formed when atoms share electrons. (Chap. 1, Sec. 2, p. 19)

N

neutralization (new truh luh ZAY shun): reaction in which an acid reacts with a base and forms water and a salt. (Chap. 3, Sec. 3, p. 84)

O

organic compounds: most compounds that contain carbon. (Chap. 4, Sec. 1, p. 96)

English Glossary

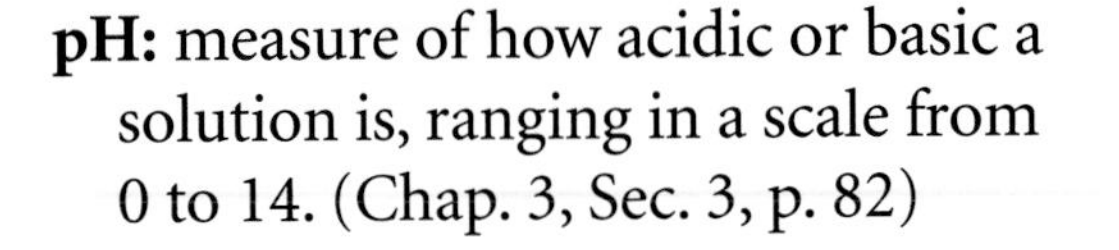

P

pH: measure of how acidic or basic a solution is, ranging in a scale from 0 to 14. (Chap. 3, Sec. 3, p. 82)

polar bond: bond resulting from the unequal sharing of electrons. (Chap. 1, Sec. 2, p. 20)

polymer: large molecule made up of small repeating units linked by covalent bonds to form a long chain. (Chap. 4, Sec. 3, p. 108)

precipitate: solid that comes back out of its solution because of a chemical reaction or physical change. (Chap. 3, Sec. 1, p. 66)

product: substance that forms as a result of a chemical reaction. (Chap. 2, Sec. 1, p. 38)

protein: biological polymer made up of amino acids; catalyzes many cell reactions and provides structural materials for many parts of the body. (Chap. 4, Sec. 3, p. 109)

R

rate of reaction: measure of how fast a chemical reaction occurs. (Chap. 2, Sec. 2, p. 48)

reactant: substance that exists before a chemical reaction begins. (Chap. 2, Sec. 1, p. 38)

S

saturated: describes a solution that holds the total amount of solute that it can hold under given conditions. (Chap. 3, Sec. 2, p. 74)

saturated hydrocarbon: hydrocarbon, such as methane, with only single bonds. (Chap. 4, Sec. 1, p. 98)

solubility (sahl yuh BIH luh tee): measure of how much solute can be dissolved in a certain amount of solvent. (Chap. 3, Sec. 2, p. 73)

solute: substance that dissolves and seems to disappear into another substance. (Chap. 3, Sec. 1, p. 66)

solution: homogeneous mixture whose elements and/or compounds are evenly mixed at the molecular level but are not bonded together. (Chap. 3, Sec. 1, p. 66)

solvent: substance that dissolves the solute. (Chap. 3, Sec. 1, p. 66)

substance: matter with a fixed composition whose identity can be changed by chemical processes but not by ordinary physical processes. (Chap. 3, Sec. 1, p. 64)

U

unsaturated hydrocarbon: hydrocarbon, such as ethylene, with one or more double or triple bonds. (Chap. 4, Sec. 1, p. 99)

Spanish Glossary

Este glosario define cada término clave que aparece en negrillas en el texto. También muestra el capítulo, la sección y el número de página en donde se usa dicho término.

A

acid / ácido: sustancia que libera iones H+ y produce iones hidronio cuando se disuelve en agua. (Cap. 3, Sec. 3, pág. 78)

activation energy / energía de activación: cantidad mínima de energía que se necesita para iniciar una reacción química. (Cap. 2, Sec. 2, pág. 47)

amino group / grupo amino: consta de un átomo de nitrógeno unido covalentemente a dos átomos de hidrógeno; se representa con la fórmula –NH2. (Cap. 4, Sec. 2, pág. 105)

amino acids / aminoácidos: unidades constitutivas de las proteínas; contienen tanto un grupo amino como un grupo de ácido carboxílico reemplazando los hidrógenos en un átomo de carbono. (Cap. 4, Sec. 2, pág. 105)

aqueous / acuosa: tipo de solución en que el agua es el disolvente. (Cap. 3, Sec. 2, pág. 70)

B

base / base: sustancia que acepta iones H+ y produce iones hidroxido cuando se disuelve en agua. (Cap. 3, Sec. 3, pág. 81)

C

carbohydrates / carbohidratos: compuestos orgánicos que sólo contienen carbono, hidrógeno y oxígeno; por ejemplo, los almidones, la celulosa, el glucógeno y los azúcares. (Cap. 4, Sec. 3, pág. 110)

carboxyl group / grupo carboxílico: consta de un átomo de carbono, dos átomos de oxígeno y un átomo de hidrógeno; se presenta con la formula –COOH. (Cap. 4, Sec. 2, pág. 105)

catalyst / catalizador: sustancia que acelera una reacción química pero que, al final del proceso, ella misma no se agota o altera permanentemente. (Cap. 4, Sec. 2, pág. 51)

chemical bond / enlace químico: fuerza que mantiene unidos dos átomos. (Cap. 1, Sec. 1, pág. 15)

chemical equation / ecuación química: forma abreviada para escribir los reactivos que se usan y los productos que se forman en una reacción química; a veces, muestra si se produce o absorbe energía. (Cap. 2, Sec. 1, pág. 38)

chemical formula / fórmula química: combinación de símbolos químicos y números que indica el tipo y cantidad de átomos de cada elemento que tiene una molécula. (Cap. 1, Sec. 2, pág. 24)

chemical reaction / reacción química: proceso que produce cambios químicos

Spanish Glossary

y que resulta en nuevas sustancias cuyas propiedades son diferentes a aquéllas de las sustancias originales. (Cap. 2, Sec. 1, pág. 36)

compound / compuesto: sustancia pura que contiene dos o más elementos. (Cap. 1, Sec. 2, pág. 17)

concentration / concentración: describe la cantidad de soluto presente en una solución en comparación con la cantidad de disolvente. (Cap. 3, Sec. 2, pág. 75)

covalent bond / enlace covalente: enlace químico que se forma cuando los átomos comparten electrones. (Cap. 1, Sec. 2, pág. 19)

E

electron cloud / nube electrónica: área donde los electrones con carga negativa, acomodados en niveles de energía, viajan alrededor del átomo del núcleo. (Cap. 1, Sec. 1, pág. 8)

electron dot diagram / diagrama de puntos electrónicos: símbolo químico de un elemento, el cual se rodea con una cantidad igual de puntos como electrones tiene el elemento en su nivel de energía externo. (Cap. 1, Sec. 1, pág. 14)

endothermic reaction / reacción endotérmica: reacción química en la cual se absorbe energía calorífica. (Cap. 2, Sec. 1, pág. 43)

exothermic reaction / reacción exotérmica: reacción química en la cual se libera energía calorífica. (Cap. 2, Sec. 1, pág.43)

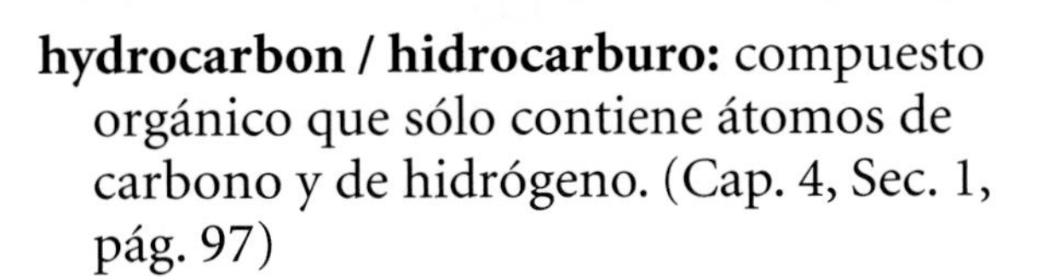

H

hydrocarbon / hidrocarburo: compuesto orgánico que sólo contiene átomos de carbono y de hidrógeno. (Cap. 4, Sec. 1, pág. 97)

hydroxyl group / grupo hidroxílico: consta de un átomo de oxígeno y uno de hidrogeno unidos por un enlace covalente; se representa con la fórmula –OH. (Cap. 4, Sec. 2, pág. 104)

I

indicator / indicador: compuesto que cambia de color a diferentes valores de pH cuando reacciona con soluciones ácidas o básicas. (Cap. 3, Sec. 3, pág. 84)

inhibitor / inhibidor: sustancia que retarda una reacción química, lo cual causa un retraso en la formación de cierta cantidad de producto. (Cap. 2, Sec. 2, pág. 50)

ion / ion: átomo que no es neutro porque ha ganado o perdido un electrón. (Cap. 1, Sec. 2, pág. 17)

ionic bond / enlace iónico: atracción que mantiene los iones con cargas opuestas cerca unos de otros. (Cap. 1, Sec. 2, pág. 17)

isomers / isómeros: compuestos que tienen la misma fórmula química, pero con distintas estructuras y distintas propiedades físicas y químicas. (Cap. 4, Sec. 1, pág. 100)

lipids / lípidos: compuestos orgánicos hechos de tres cadenas largas de ácidos carboxílicos enlazados al glicerol, las grasas, los aceites y las ceras. (Cap. 4, Sec. 3, pág. 113)

M

molecule / molécula: partícula neutra que se forma cuando los átomos comparten electrones. (Cap. 1, Sec. 2, pág. 19)

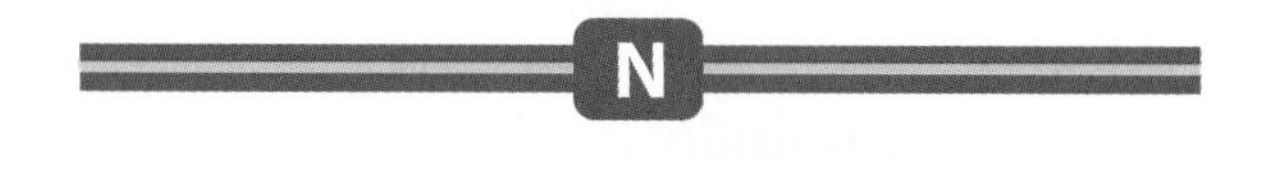

neutralization / neutralización: reacción en que un ácido reacciona con una base y se forma agua y una sal. (Cap. 3, Sec. 3, pág. 84)

O

organic compounds / compuestos orgánicos: la mayoría de los compuestos que contienen carbono. (Cap. 4, Sec. 1, pág. 96)

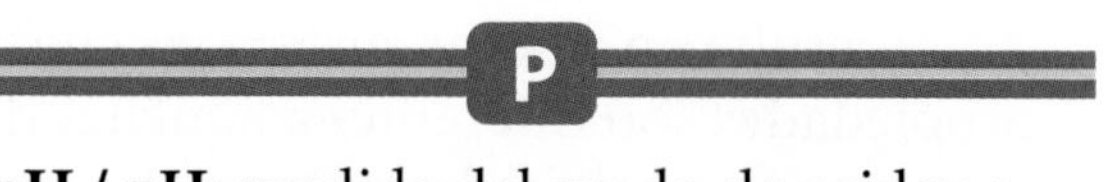

pH / pH: medida del grado de acidez o basicidad de una solución, se mide de 0 a 14 en la escala de pH. (Cap. 3, Sec. 3, pág. 82)

polar bond / enlace polar: enlace que resulta del intercambio desigual de electrones. (Cap. 1, Sec. 2, pág. 20)

polymer / polímero: molécula grande hecha de pequeñas unidades repetitivas unidas por enlaces covalentes para formar una cadena larga. (Cap. 4, Sec. 3, pág. 108)

precipitate / precipitado: sólido que debido a una reacción química o a un cambio físico que se lleva a cabo en una solución. (Cap. 3, Sec. 1, pág. 66)

product / producto: sustancia que se forma como resultado de una reacción química. (Cap. 2, Sec. 1, pág. 38)

protein / proteína: polímero biológico hecho de aminoácidos; cataliza muchas reacciones celulares y provee materiales estructurales para muchas partes del cuerpo. (Cap. 4, Sec. 3, pág. 109)

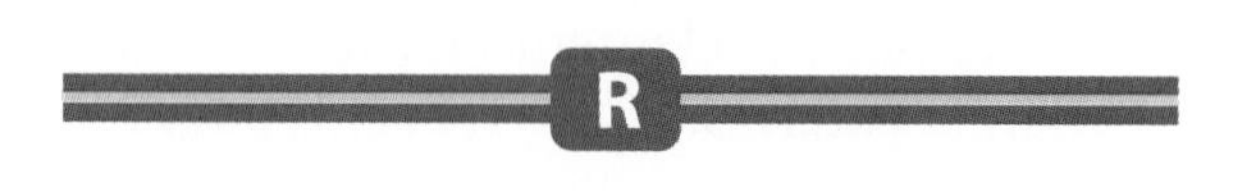

rate of reaction / velocidad de la reacción: medida del grado de velocidad con que ocurre una reacción química. (Cap. 2, Sec. 2, pág. 48)

reactant / reactivo: sustancia que existe

antes del inicio de una reacción química. (Cap. 2, Sec. 1, pág. 38)

saturated / saturada: describe una solución que tiene toda la cantidad de disolvente que puede sostener bajo ciertas condiciones. (Cap. 3, Sec. 2, pág. 74)

saturated hydrocarbon / hidrocarburo saturado: hidrocarburo, como el metano, que sólo tiene enlaces simples. (Cap. 4, Sec. 1, pág. 98)

solubility / solubilidad: medida de la cantidad de soluto que se puede disolver en cierta cantidad de disolvente. (Cap. 3, Sec. 2, pág. 73)

solute / soluto: sustancia que disuelve y parece desaparecer en otra sustancia. (Cap. 3, Sec. 1, pág. 66)

solution / solución: mezcla homogénea cuyos elementos y compuestos están distribuidos uniformemente a nivel molecular, pero que no están unidos. (Cap. 3, Sec. 1, pág. 66)

solvent / disolvente: sustancia que disuelve el soluto. (Cap. 3, Sec. 1, pág. 66)

substance / sustancia: materia con una composición fija cuya identidad se puede cambiar mediante procesos químicos, pero que no se puede cambiar a través de procesos común y corrientes o físicos. (Cap. 3, Sec. 1, pág. 64)

U

unsaturated hydrocarbon / hidrocarburo no saturado: hidrocarburo como el etileno, que tiene uno o más enlaces dobles o triples. (Cap. 4, Sec. 1, pág. 99)

Index

The index for *Chemistry* will help you locate major topics in the book quickly and easily. Each entry in the index is followed by the number of the pages on which the entry is discussed. A page number given in boldfaced type indicates the page on which that entry is defined. A page number given in italic type indicates a page on which the entry is used in an illustration or photograph. The abbreviation *act.* indicates a page on which the entry is used in an activity.

A

B

C

Index

Index

M

O

P

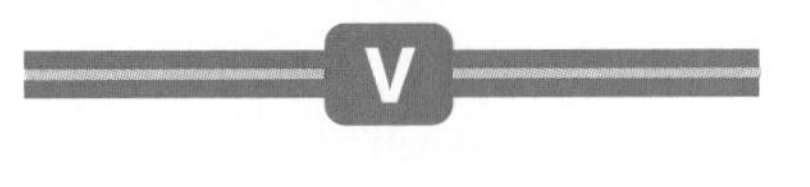

W

Art Credits

Glencoe would like to acknowledge the artists and agencies who participated in illustrating this program: Absolute Science Illustration; Andrew Evansen; Argosy; Articulate Graphics; Craig Attebery represented by Frank & Jeff Lavaty; CHK America; Gagliano Graphics; Pedro Julio Gonzalez represented by Melissa Turk & The Artist Network; Robert Hynes represented by Mendola Ltd.; Morgan Cain & Associates; JTH Illustration; Laurie O'Keefe; Matthew Pippin represented by Beranbaum Artist's Representative; Precision Graphics; Publisher's Art; Rolin Graphics, Inc.; Wendy Smith represented by Melissa Turk & The Artist Network; Kevin Torline represented by Berendsen and Associates, Inc.; WILDlife ART; Phil Wilson represented by Cliff Knecht Artist Representative; Zoo Botanica.

Photo Credits

Abbreviation Key: AA=Animals Animals; AH=Aaron Haupt; AMP=Amanita Pictures; BC=Bruce Coleman, Inc.; CB=CORBIS; DM=Doug Martin; DRK=DRK Photo; ES=Earth Scenes; FP=Fundamental Photographs; GH=Grant Heilman Photography; IC=Icon Images; KS=KS Studios; LA=Liaison Agency; MB=Mark Burnett; MM=Matt Meadows; PE=PhotoEdit; PD=PhotoDisc; PQ=PictureQuest; PR=Photo Researchers; SB=Stock Boston; TSA=Tom Stack & Associates; TSM=The Stock Market; VU=Visuals Unlimited.

Cover PD; **iv** Christopher Swann/Peter Arnold, Inc.; **v** Victoria Arocho/AP/Wide World Photos; **vi** file photo; **1** KS; **2** (t)Bettmann/CB, (bl)Hulton Getty/Archive Photos, (br)SuperStock; **3** (t)Grant V. Faint/The Image Bank, (b)Gianni Dagli Orti/CB; **4** (t)Lynn Eodice/IndexStock, (b)Fred Charles/Stone; **5** AP/Wide World Photos; **6** IBMRL/VU; **6-7** Paul M. Walsh/The Morning Journal/AP/Wide World Photos; **7** AH; **15** Laura Sifferlin; **16** (l)Lester V. Bergman/CB, (r)DM; **21** MM; **22** (t, l to r)PhotoTake NYC/PQ, Albert J. Copley/VU, Kenneth Libbrecht/Caltech, Kenneth Libbrecht/Caltech, (b)Manfred Kage/Peter Arnold, Inc.; **24** James L. Amos/PR; **25 26 27** AH; **28** Fulcrum Publishing; **29** Tobias Rostlund/AP/Wide World Photos; **34** Chuck Pefley/SB; **34-35** Lester Lefkowitz/TSM; **35** MM; **36** (l)AH, (r)DM; **37** (tl)Patricia Lanza, (tc)Jeff J. Daly/VU, (tr)Susan T. McElhinney, (bl)Craig Fujii/Seattle Times, (br)Sovfoto /Eastfoto/PQ; **38** AP/Wide World Photos; **41** Sovfoto/Eastfoto/PQ; **43** Christopher Swann/Peter Arnold, Inc.; **44** (t)AH, (c)Frank Balthis, (b)Tom Stewart/TSM; **45** David Young-Wolff/PE/PQ; **46** (l)AP, (r)Richard Megna/FP; **47** (l)Victoria Arocho/AP/Wide World Photos, (r)Skip Nall/PD; **48** (t bl)AH, (br)IC; **49** SuperStock; **50** (l)Chris Arend/Alaska Stock Images/PQ, (r)AH; **51** Courtesy General Motors; **52** MM; **53** Timothy Fuller; **54** (t)AP, (b)Bob Daemmrich; **56** Tino Hammid Photography; **57** Joe Richard/UF News & Public Affairs; **58** (t)AH, (c)MM, (b)Spencer Grant/PE; **59** David Young-Wolff/PE; **62** SuperStock; **62-63** Ed Crockett; **63** John Evans; **65** (l)Stephen W. Frisch/SB, (r)DM; **66** (t)Ray Pfortner/Peter Arnold, Inc., (b)DM; **67** Richard Hamilton/CB; **68** John Evans; **69** (l)SuperStock, (r)Annie Griffiths/CB; **72** John Evans; **74** Richard Nowitz/Phototake NYC/PQ; **76** (t)Claire Paxton & Jacqui Farrow/Science Photo Library/PR, (b)AH; **79** John Evans; **80** (tl)Joe Sohm/Chromosohm/Stock Connection/PQ, (tc)Andrew Popper/Phototake NYC/PQ, (tr)A. Wolf/Explorer/PR, (b)Stephen R. Wagner; **81** John Evans; **86** (t)Rick Weber, (b)KS; **87** KS; **88** (t)CB, (b)Myrleen Ferguson Cate/PE; **90** (t)John Evans, (c)John A. Rizzo/PD, (b)L.S. Stepanowicz/Panographics; **92** MM; **94** IC; **94-95** Index Stock; **95** MM; **96** (l)Michael Newman/PE, (r)Richard Price/FPG; **98** (l)Tony Freeman/PE, (r)MB; **99** (l c)MB, (r)Will & Deni McIntyre/PR; **100** Ted Horowitz/TSM; **103** KS; **105** (t)Kim Taylor/BC/PQ, (b)John Sims/Stone; **107** KS; **109** AH; **110** (t)Elaine Shay, (b)Mitch Hrdlicka/PD; **111** (t)KS, (b)MM; **112 113** KS; **114** (l)Don Farrall/PD, (r)KS; **115** Alfred Pasieka/Peter Arnold, Inc.; **116** (t)Geoff Butler, (b)AH; **117** AH; **118** (t)David Nunuk/Science Photo Library/PR, (bl)Lynne Johnson/Aurora, (br)Waina Cheng/BC; **119** (l)Lee Baltermoal/FPG, (r)Richard Johnston/FPG; **119** (inset)Waina Cheng/BC; **120** Hans Pfletschinger/Peter Arnold, Inc.; **121** AH; **122** CB; **126-127** PD; **128** (t)KS, (b)Bill Aron/PR; **129** (t)IC, (bl br)KS; **130** (t)Don Tremain/PD, (b)AH; **131** KS; **132** Timothy Fuller; **136** Roger Ball/TSM; **138** (l)Geoff Butler, (r)Coco McCoy/Rainbow/PQ; **139** Dominic Oldershaw; **140** StudiOhio; **141** First Image; **143** MM; **146** Paul Barton/TSM; **149** Davis Barber/PE.

MARK TWAIN

LAS AVENTURAS DE HUCK FINN

EDIVAL - ALFREDO ORTELLS

Título original: *The Adventures of Huckleberry Finn*
Traductor: José-Félix
Introducción: José-Félix

ISBN 84-7404-015-9
ISBN 84-7404-014-0
ISBN 84-7189-029-1
ISBN 84-7189-030-5
Depósito legal: M. 36.280-1975

Papel fabricado por Torras Hostench, S. A.
Fotocomposición, Compoprint (Marqués de Monteagudo, 16 - Madrid)
Impreso en Mateu-Cromo (carretera Pinto a Fuenlabrada, s/n - PINTO, Madrid) - Printed in Spain

MARK TWAIN

En busca de su estrella

Samuel Langhorne Clemens, a quien se conocerá con el pseudónimo de «Mark Twain», nació un 30 de noviembre de 1835 en un pueblo llamado Florida del estado de Missouri. Su familia era del Sur, patricia, es decir descendiente de primitivos pobladores ingleses, que conservaban el orgullo de su abolengo. Su padre, John Marshall Clemens, de Virginia, ejercía la carrera de abogado. Quiso hacer fortuna en los negocios, pero sin suerte, por lo que estuvo siempre al borde de la ruina. Su madre, Jane Lampton, de Kentucky, fue una mujer refinada y ejerció una gran influencia en el futuro escritor. Siendo niño, los fracasos comerciales de su padre llevaron la familia a Hannibal, en el límite con el estado de Illinois.

Aquí pasó su niñez Clemens, en una región que era por entonces el centro geográfico de Estados Unidos, una naturaleza sin explorar a la que llegaban pioneros en busca de tierras de cultivo, explotaciones de madera y minerales. Por el Mississippi bajaban aventureros con sus ilusiones a cuestas. El escenario de la mejor obra de Mark Twain, así como lo más sustancioso de su carácter, su sentido del humor, su sátira, su filosofía provienen de las granjas, de los bosques, del río y de la gente que vivía a orillas del Mississippi.

El pequeño Clemens iba a la escuela en Hannibal. No le gustaba mucho y sus años escolares son cortos. Sin embargo devoraba los libros de aventuras de una pequeña librería del pueblo.

Murió su padre cuando tenía doce años. La familia quedó en la miseria y Clemens tuvo que dejar la escuela para empezar a trabajar. Un tío suyo, establecido en Hannibal, tenía una pequeña imprenta en la que editaba un periódico, que más tarde dirigiría el hermano mayor de Clemens, Orion. Aquí aprendió el oficio de cajista *y pronto, por entretenimiento, escribió articulillos humorísticos, crónicas, «sketches», que divertían a los amigos. Uno de los primeros artículos en el «Carpet Bag» de Boston parece que fue* The Dandy Frightening the Squatter (El petrimetre que asusta al usurpador, *1852), firmado «S.L.C. de Hannibal».*

A los dieciocho años salió de casa para recorrer mundo y trabajó de cajista en St. Louis, Nueva York, Philadelphia. Fue a Iowa, donde se había trasladado su hermano Orion. Allí escribió una serie de cartas humorísticas para el «Keokuk Saturday Post» con el pseudónimo de «Thomas Jefferson Snodgrass». Ya despuntaba su inclinación hacia el humorismo.

Una de las ilusiones de su juventud fue ser piloto de los barcos que cruzaban el Mississippi. Quiso viajar a América del Sur. Inició el viaje, pero durante la travesía encontró a un piloto del que se hizo amigo y discípulo. De esta forma se quedó dos años en el Mississippi como aprendiz de piloto fluvial. Clemens aprendió todos los secretos de la navegación y los recodos del gran río y también allí aprendió, en los pueblos y ciudades de la orilla y en los viajeros que transportó, a conocer los más variados aspectos de la naturaleza humana.

Encontró un rumbo

Quizás se habría quedado Clemens en el Mississippi como piloto, si la guerra civil de 1861 no hubiera interrumpido su carrera (El Norte enarbolaba la bandera del antiesclavismo contra el Sur, cuya riqueza agrícola se basaba en la esclavitud de los negros). Se alistó en un grupo de voluntarios del Sur, pero se disolvió éste antes de prestar juramento en el ejército confederado. Se marchó al Oeste tras su hermano Orion, nombrado secretario del gobernador de Nevada. Se contagió de la fiebre de los pioneros del Oeste y pensó hacer grandes negocios: buscó minas de plata, especuló con tierras, probó negocios de maderas, pero fracasó. Mientras tanto seguía enviando sus crónicas a los periódicos. El «Territorial Enterprise» le ofreció el empleo de informador local en Virginia City. Aquí escogió su pseudónimo de «Mark Twain» (era el grito de los sondeadores del Mississippi. «Mark twain!», dos brazas de profundidad, o sea profundidad suficiente, navegación sin riesgos). Dada su inclinación a la sátira tuvo que dejar el territorio de Nevada por un conato de duelo con el director de un periódico de Virginia.

Dimite como informador y se va a los que fueron grandes campos auríferos de California. Aquí oyó contar el relato que le haría famoso, La célebre rana saltarina del condado de Calaveras (The Celebrated Jumping Frog of Calaveras County), *publicado en el «Saturday Press» en 1865.*

Vuelve a San Francisco sin un real y trabaja como reportero del «Call», corresponsal del «Territorial Enterprise» y colaborador de «The Golden Era» y «The Californian».

A raíz del éxito de La rana saltarina... *el «Sacramento Union» le encomienda un reportaje sobre las islas Sandwiche (más tarde Hawaii) y escribió unas cartas con su enfoque humorístico. Por entonces abordó uno de sus temas favoritos: el fracaso moral de la mal llamada civilización. La razón, el progreso, la técnica de la civilización occidental destruían el paraíso original.*

Acuciado por la necesidad toca su tecla de oro: el humorismo. Se presenta en los teatros dando conferencias humorísticas y tiene un éxito arrollador, que le proporcionará siempre el dinero que busque.

Se organiza en 1866 una excursión a Europa y Tierra Santa a bordo del «Quaker City» y el diario «Alta California» le comisiona que envíe relaciones epistolares del viaje. Tuvieron gran éxito sus reportajes, recogidos más tarde en Gente ingenua en el extranjero (Innocents

Abroad, *1869). Mark Twain describe las reacciones del norteamericano tipo, ingenuo, pragmático, primitivo y sensato ante la complejidad de la cultura europea y oriental. La obra resulta pobre en conjunto por la magnitud del mundo a describir. Además el autor ni pudo ni quiso entender las obras de arte y menos las sutilezas, complejidades y recovecos espirituales del Viejo Mundo. Twain no podía comprender el aspecto estético, no práctico de la vida, la belleza de los templos en el centro de la miseria de los pueblos católicos.*

Su gran amor

Al regresar de este viaje, se coloca de secretario del senador de Nevada y escribe para varios periódicos Cartas desde la Colina del Capitolio. *Había conocido durante la excursión a Europa a Charles Langdon, hijo de una rica familia de Elmira. La hermana de éste, Olivia, mujer cariñosa, rígida en cuestiones morales y enfermiza, se casó con Mark Twain el 2 de febrero de 1870. Fue un matrimonio feliz.*

Con el dinero de sus conferencias compró una participación en el diario «Express» de Buffalo y allí se establecieron los recién casados en una bonita casa, obsequio del padre de Olivia. En Buffalo nació su hijo Langdon, muerto poco después. Abandona la redacción del periódico, pues no era un hombre de horario. En 1872 se trasladan a Hartford, en Connecticut, sede fija del matrimonio. Publica Roughing It (Vida dura, *también traducida* Pasando fatigas, *1872), aprovechando sus recuerdos del lejano Oeste, donde el «abogado», el editor, el banquero, el más desesperado, el mayor jugador y el cantinero ocupaban el mismo lugar en la sociedad. Nace su hija Susy. Embarca para Inglaterra y la conquista con su humor y sigue ganando dinero con sus conferencias en Estados Unidos.*

En colaboración con Charles Dudley Warner publica en 1873 su primera novela larga, La edad dorada (The Gilded Age) *(Es una pintura de la sociedad americana durante los años en que lanzó a sus hombres hacia el Oeste para poblarlo, mecanizarlo y acometer empresas gigantescas).*

Nace su hija Clara en 1874 y empieza Las aventuras de Tom Sawyer (The Adventures of Tom Sawyer), *que publica al año siguiente. Con esta obra adquiere fama de escritor extraordinario.*

Animado por el éxito de Las aventuras de Tom Sawyer, *se puso a escribir la obra que, en opinión de muchos, es la mejor,* Las aventuras de Huckleberry Finn (The Adventures of Huckleberry Finn), *pero dejó el manuscrito y no volvió a tomarlo hasta 1884 tras un viaje por el Mississippi para refrescar los recuerdos de su niñez y de su juventud.*

Vasta epopeya de la América de los aventureros y de los miserables ciudadanos diseminados a lo largo del valle del Missouri y de Ohio, la América de la edad de oro y de la colonización, de la vida violenta y elemental. Huck es el retrato del «boy» americano de aquellos tiempos.

Huck es el desventurado hijo de un padre indigno y borracho. Abandonado, viene acogido por personas buenas, que hubieran deseado ocuparse de su educación, pero su padre (atraído por el dinero que Huck y Tom encontraron en la cueva al final de *Las aventuras de Tom Sawyer)* se lo lleva a una cabaña perdida entre los bosques. El muchacho consigue escaparse y arriba a una isla, en la que está refugiado Jim, un negro y viejo amigo de los muchachos, amenazado con ser vendido en el mercado. Para evitar que encuentren al negro, dado que su rescate tiene un precio, navegan de noche en una balsa tratando de llegar a un Estado abolicionista, para que Jim recobre su libertad. Durante un fuerte temporal pierden el rumbo y se topan con dos embrollones, a los que se ven obligados a subir a la balsa. Estos, tras diversas aventuras para sacar dinero a la pobre gente de la ribera, venden a traición al negro Jim a un tío de Tom Sawyer. Fortuito encuentro de los dos muchachos, con gran sorpresa de Tom, que creía muerto a Huck. Aventuras fantásticas para una fantasiosa liberación de Jim. Aclarada la verdadera identidad de los muchachos y el motivo de la liberación del negro, el libro termina con la perspectiva de la escuela para Huck y su alusión a cierto plan de fugarse con los indios.

Publica en el «Atlantic Monthly» una de sus obras-recuerdo de los años felices de su niñez, Viejos tiempos en el Mississippi (Old Times on the Mississippi, *1875).*

Nace su última hija, Jean, y publica Un vagabundo en el extranjero (A Tramp Abroad, *1880). Al año siguiente se publica en Londres* El príncipe y el mendigo (The Prince and the Pauper).

Durante el viaje por el Mississippi para rememorar el material de su proyectado Huckleberry Finn, *saca tema para una de sus más vivaces obras sobre la tierra de sus primeros años,* La vida en el Mississippi (Life on the Mississippi, *1883).*

Nueva gira de conferencias con éxito espectacular. Durante la gira conoce la obra de Malory, La muerte de Arturo *(especie de gesta medieval). Le entusiasmó y de aquí nació una obra maestra,* Un yanqui de Connecticut en la corte del rey Arturo (A Connecticut Yankee at King's Arthur Court, *1889).*

La celebridad le trajo una prosperidad económica. Se embarcó en dos negocios: fundó la editorial «Webster» y financió un proyecto de una máquina de componer, una especie de linotipia. Los gastos en uno y otro negocio juntamente con la crisis financiera de 1890-1893 dejaron a Mark Twain en la ruina. Gracias al asesoramiento de su admirador, Henry H. Rogers, de la Standard Oil Company, resolvió su problema financiero.

Viaje a Europa en 1893. Se dedica a una frenética producción, Novela de una joven esquimal, *publicada en la revista «Cosmopolita»,* El calabaza Wilson (Pudd'n Head Wilson, *1894) en «Century»* y Tom Sawyer en el extranjero *en «St. Nicholas Magazine». Realiza una gira de conferencias humorísticas por el mundo anglosajón: Australia, Nueva Zelanda, Asia, Sudáfrica. Exito arrollador y dinero para pagar todas las deudas. Las notas de este viaje le sirvieron para uno de sus libros de viaje más importantes,* En torno al Ecuador (Following the Equator, *1897) (Es una crítica violenta y pesimista del imperialismo. Relata el tráfico de esclavos en las islas del Pacífico, el exterminio de los kanakas, la explotación de los nativos por parte de los blancos* civilizados).

Sátira grotesca

La familia vivió en Europa los años 1896-1900.

La muerte de su hija mayor Susy (1896), las muertes de su hermano Orion y de su hermana Pamela y los primeros ataques epilépticos de su hija Jean hicieron perder su humor alegre. Siguió creyendo en la vida, pero el humor optimista y fresco de sus primeras obras cedió a un humor amargo, de sátira grotesca. Apareció una de las obras que más años de reflexión costó al autor y que él consideraba la mejor salida de su pluma, Recuerdos personales de Juana de Arco (Personal Recollections of Joan of Arc, *1896).*

Regresa a Estados Unidos en 1900 y se le acoge como un acontecimiento nacional. Las universidades de Yale, Missouri y más tarde Oxford le conceden el título de «doctor honorario». Enferma su esposa y el médico le recomienda un cambio de clima. Vuelven a Europa y se instalan en Florencia, donde Olivia muere el 5 de junio de 1904. Mark Twain lleva el cadáver del ser que más ha amado en la vida para que descanse en la tumba familiar de Elmira. El escritor no se repondrá de la pérdida de su compañera, nunca se resignó al vacío que dejaba después de treinta y siete años de felicidad. Sus últimos libros son muy pesimistas. Sentía que en el fondo del hombre anida desde el principio el egoísmo y la crueldad. En El misterioso forastero (The Mysterious Stranger, *1906, y publicado póstumamente en 1913) resume su pesimismo: «la vida toda es un sueño, un sueño absurdo». Entre las obras que podemos destacar en su final, encontramos* Mi novia platónica *(revelación del subconsciente del escritor) y el* Diario de Eva, *en el que parece esperar lo que él llamaba su liberación (la muerte).*

Murió en Reading, Connecticut, el 21 de abril de 1910. Abrió los ojos, agarró la mano de su única hija viva, Clara, y le miró a la cara: «¡Adiós, cariño! Si acaso nos encontramos...»

El paraíso perdido

Mark Twain fue un humorista que supo canalizar el gusto y la filosofía de la vida de sus compatriotas de la primera mitad del siglo XIX. *Cristalizó en sus obras el estilo de vida de los Estados Unidos hasta la Guerra Civil. En sus mejores obras se manifiestan la ingenuidad, el optimismo de los pioneros, el afán de libertad, el amor a la naturaleza, el sentido práctico e igualitario, la desconfianza ante el poder establecido y la socarronería, superstición e ignorancia del campesino. Frente a las ciudades del Este, en las que se iba acumulando la cultura antigua y la civilización industrial, el Oeste de Mark Twain es el símbolo de la libertad y la inocencia. El tema central de su obra es* el paraíso perdido.

La obra de Mark Twain, distribuida en un arco de cuarenta y cinco años, tiene dos vertientes:

1) En la etapa en que escribe sus obras más difundidas, Las aventuras de Tom Sawyer *(1875),* Las aventuras de Huckleberry Finn *(1884),* La vida del Mississippi *(1883), su humorismo se proyecta sobre un fondo de afirmación de la vida. La libertad y la magia son los motivos esenciales. Tom Sawyer quiere ser libre de jugar, amar, arriesgarse, ser hombre. Huckleberry Finn, el vagabundo, es más libre que Tom. Será su amigo y su maestro en la vida. Pero viven en un mundo de misterio, zonas inexploradas están sujetas a la superstición. Esta etapa está caracterizada por el canto a la naturaleza, por la profunda afinidad que existe entre el hombre y el bosque.*

2) El aspecto más maduro de sus obras, aunque de menor calidad y menos conocidas por el gran público, es la segunda mitad de su vida. Es la época de la sátira social, de la denuncia de las hipocresías que anidaban en las más respetables instituicones. Mark Twain criticó el imperialismo de las grandes potencias en En torno al Ecuador: *represión norteamericana en Filipinas, las atrocidades belgas en el Congo, el despotismo zarista, la civilización cristiana, la guerra de Inglaterra contra los boers, la represión de los boxers en China. La mayoría de los artículos están en* A quien se sienta en la oscuridad (To the Person Sitting in Darkness), El soliloquio del rey Leopoldo, El soliloquio del zar. *Al final de su vida la sátira social, la protesta política de Mark Twain se resumían en una condena total del género humano: se había perdido la inocencia primordial, el paraíso.*

JOSÉ-FÉLIX

AVISO

Las personas que intenten encontrar un motivo de este relato, serán procesadas; las personas que traten de encontrarle una moraleja, serán desterradas; las personas que traten de encontrarle un argumento, serán fusiladas.

Por orden del autor

G. G.
Jefe de artillería

1

No saben quién soy si no han leído un libro titulado *Las aventuras de Tom Sawyer,* pero no importa. Ese libro lo escribió Mark Twain y en líneas generales dijo la verdad. Algunas cosas las exageró, pero en conjunto dijo la verdad. Eso no es nada. Jamás he visto a nadie que no mienta alguna que otra vez, exceptuando a tía Polly, o la viuda, e incluso Mary. Tía Polly —la tía Polly de Tom—, Mary y la viuda Douglas, de las que habla ese libro, que en conjunto es un libro sincero con algunas exageraciones, como ya dije antes.

El libro termina de esta manera: Tom y yo encontramos el dinero que los ladrones escondieron en la cueva y nos volvimos ricos. Nos correspondieron seis mil dólares a cada uno en oro. Daba miedo ver tanto dinero amontonado. Bueno, pues el juez Thatcher lo puso a interés y nos daba un dólar diario a cada uno durante todo el año... más de lo que uno puede hacer con él. La viuda Douglas me adoptó como si fuera hijo suyo y afirmó que me civilizaría, pero era duro vivir siempre en casa, teniendo en cuenta las costumbres tan regulares y decentes que tenía la viuda. Así que, cuando no pude aguantar más, me escabullí. Me puse mis viejos andrajos, me metí en mi barril de azúcar otra vez y me encontré libre y satisfecho. Pero Tom Sawyer fue en mi busca y dijo que pensaba organizar una banda de ladrones y que podría unirme a ellos si volvía al lado de la viuda y era respetable. De manera que volví.

La viuda lloró mucho por mí y me llamó «pobre corderito extraviado» y muchas otras cosas, pero, eso sí, sin ninguna mala intención. Me puso otra vez trajes nuevos y yo no hice más que sudar y sudar y notarme encogido. Bueno, comenzaba otra vez lo de antes. La viuda pedía la cena tocando una campanilla y uno tenía que presentarse con puntualidad. Cuando llegaba a la mesa, no empezaba a comer en seguida, sino que tenía que esperar a que la viuda bajara la cabeza y gruñera un poco encima de la comida, aunque ésta no tenía nada malo. Es decir, solamente que estaba guisado todo aparte. Es distinto cuando se echa todo junto dentro de una cazuela y se mezcla, uniendo el jugo de unas cosas con otras, y así tiene mejor sabor.

Después de cenar sacó un libro y me habló de Moisés y los juncos, y me entraban sudores para entenderla, pero poco a poco ella me hizo comprender que Moisés había muerto hacía muchísimo tiempo, de modo que dejé de interesarme por él, porque los muertos no me hacen mucha gracia.

Tuve ganas de fumar y pedí permiso a la viuda, pero se negó. Dijo que era una fea costumbre y poco limpia, y que debía intentar dejar ese vicio. Hay personas así. Hablan mal de una cosa cuando no saben nada de ella. Ahí estaba ella, preocupándose de Moisés, que no era nada suyo ni útil para nadie, ya que había muerto, y, sin embargo, a mí me criticaba hacer algo que era agradable. Tomaba rapé, pero esto estaba bien, porque lo hacía ella, claro...

Su hermana, la señorita Watson, una solterona flaca y soportable, con gafas, había venido recientemente a vivir con ella y la tomó conmigo persiguiéndome con una gramática. Me hacía trabajar de lo lindo hora tras hora hasta que la viuda la obligó a aflojar las riendas. No habría podido soportarlo mucho más tiempo. Después fue mortalmente aburrido durante una hora y me puse nervioso. La señorita Watson decía: «Los pies no se ponen aquí encima, Huckleberry», y: «Huckleberry, ponte derecho; siéntate bien», y, poco después: «No bosteces ni te despereces así, Huckleberry... ¿Por qué no tienes más compostura?» Después me habló del infierno y yo dije que ojalá yo estuviera allí. Se puso como loca, pero no fue mi intención enojarla. Lo que yo quería era ir a cualquier parte; sólo deseaba cambiar, no quería ningún sitio en particular. Dijo que era muy malo por decir aquello, que ella no lo diría por nada del mundo y que viviría como era debido para ir al cielo. Bueno, no encontré ninguna ventaja en ir a donde iba ella, de modo que decidí no imitarla. Pero no lo dije, porque solamente habría buscado conflictos inútiles.

Había arrancado a hablar y siguió contándome cosas del cielo. Dijo que allí arriba lo único que hace uno es pasearse todo el día con un arpa y estar cantando siempre. No me agradó mucho la idea. Pero no lo dije. Le pregunté si creía que Tom Sawyer iría al cielo, y ella dijo que ni soñarlo... Me alegré, porque quería que Tom y yo estuviéramos juntos.

La señorita Watson continuó fastidiándome, y yo me cansé y empecé a sentirme solo. Luego fueron en busca de los negros, rezaron las oraciones y todos se acostaron. Subí a mi cuarto con un pedazo de vela y lo puse encima de la mesa. Me senté en una silla junto a la ventana y traté de pensar en algo divertido, pero fue inútil. Estaba tan sólo, que deseé morir. Brillaban las estrellas y el rumor de las hojas en el bosque era melancólico, y oí el distante ulular de un búho anunciando la muerte de alguien, y un chotacabras y un perro aullando por alguien que iba a morir, y el viento que trataba de susurrarme algo, sin que yo lo entendiera, y me entraron escalofríos. Luego, allá a los lejos, en los bosques, oí esa especie de sonido que hace un fantasma cuando quiere decir algo que tiene en el pensamiento y no se hace entender y por ello se remueve en su tumba y tiene que vagar todas las noches gimiendo.

Me sentí tan alicaído y asustado, que deseé tener compañía. Entonces encontré una araña sobre mi hombro y, al sacudírmela, cayó encima de la vela y, antes de que yo pudiera mover un dedo, ardió por completo. No era preciso que nadie me dijera que aquello era un mal presagio, que me acarrearía mala suerte. Estaba completamente despavorido y tembloroso. Me levanté y di tres vueltas haciéndome una cruz en el pecho cada vez; después me até un mechón de pelos con un hilo para mantener a raya a las brujas. Pero no tenía confianza. Esto se hace cuando se pierde una herradura que no se ha encontrado, en vez de clavarla en lo alto de la puerta, pero nunca oí decir a nadie que fuera un medio seguro de ahuyentar la mala suerte cuando uno ha matado una araña.

Me senté otra vez, temblando de pies a cabeza, y saqué la pipa para fumar, porque la casa estaba tan quieta como la muerte y la viuda no se enteraría. Bueno, al cabo de largo rato oí el lejano reloj del pueblo que retumbaba doce veces... Después todo quedó silencioso de nuevo, más que antes. A poco oí el crujido de una ramita abajo, en la oscuridad, entre los árboles... Algo se movía. Permanecí sentado, quieto, escuchando. Apenas pude oír un «¡Miaaaau! ¡Miaaaau!» abajo. ¡Estupendo! Yo respondí: «¡Miaaaau! ¡Miaaaau!», tan bajito como pude, y luego apagué la luz y me descolgué por la ventana sobre el cobertizo. De allí salté al suelo y me arrastré entre los árboles y, en efecto, allí estaba Tom Sawyer esperándome.

2

Anduvimos de puntillas por un sendero entre los árboles, hacia la salida del jardín de la viuda, agachados para que las ramas no nos dieran en las cabezas. Cuando pasamos junto a la cocina, caí sobre una raíz e hice ruido. Nos agazapamos y estuvimos quietos. El corpulento negro de la señorita Watson, que se llamaba Jim, estaba sentado a la puerta de la cocina. Podíamos verlo perfectamente porque tenía la luz detrás de él. Se puso de pie y alargó el cuello un minuto, escuchando. Luego dijo:

—¿Quién hay por ahí?

Siguió escuchando; luego se acercó de puntillas y se quedó entre nosotros; casi podíamos tocarlo. Bueno, lo más seguro era que pasaran minutos y más minutos antes de que hubiera otro ruido, ya que estábamos tan cerca los unos de los otros. En el tobillo me entró un picor, pero no me atreví a rascarme. Luego empezó a picarme la oreja, después la espalda, justamente entre los hombros. Parecía que iba a morirme si no me arrascaba. Bueno, desde entonces he notado esto montones de veces. Si uno está entre gente bien, o en un entierro, o trata de dormir sin tener sueño... si uno está en cualquier parte donde no está bien visto rascarse, ¿por qué ha de picarnos todo de repente? Jim dijo después:

—¿Quién es? ¿Dónde está? ¡Maldita sea, yo he oído algo! Bien, ya sé lo que he de hacer. Voy a sentarme y estaré escuchando hasta que vuelva a oírlo.

Y se sentó en el suelo, entre Tom y yo. Apoyó la espalda contra un árbol y estiró las piernas hasta que una de las suyas casi tocó la mía. Empezó a picarme la nariz. Me picó hasta que me lloraron los ojos. Pero no me atreví a rascarme. Después me picó por debajo. No sabía cómo iba a poder estarme quieto. Este tormento duró una eternidad de seis o siete minutos. Después empezaron a picarme once sitios distintos. Pensé que no podría soportarlo un minuto más, pero apreté los dientes y traté de conseguirlo. Entonces Jim empezó a respirar pesadamente, después a roncar... y yo me encontré en seguida estupendamente.

Tom me hizo una seña, una especie de ruido con la boca, y nos alejamos a gatas. Cuando habíamos avanzado unos diez pies, Tom me susurró al oído que quería atar a Jim al árbol para divertirnos, pero yo dije que no. Podía despertarse y armaría un jaleo gordo, y entonces descubrirían que yo no estaba en mi cuarto. Tom dijo que no teníamos velas suficientes y que entraría en la cocina para coger algunas. No quería que él lo intentara. Dije que Jim se despertaría. Pero Tom quiso correr el riesgo, de manera que entramos en la cocina y cogimos tres velas, y Tom las pagó dejando cinco centavos encima de la mesa. Después salimos y yo sudaba de ganas de marcharme, pero Tom estaba empeñado en acercarse a gatas a Jim y hacerle algo. Esperé durante un buen rato, o así me pareció. Estaba todo tan silencioso y solitario...

En cuanto volvió Tom, nos fuimos por el sendero, dimos la vuelta a la valla del jardín y a poco llegamos a la cumbre escarpada de una colina al otro lado de la casa. Tom dijo que había quitado el sombrero a Jim colgándolo de una rama que había encima de su cabeza, que Jim se había movido ligeramente, pero no se despertó. Jim dijo después que las brujas le habían echado un maleficio y cabalgado sobre él por todo el Estado, y que le dejaron de nuevo debajo de los árboles colgando el sombrero en una rama para demostrarle que habían estado allí. La segunda vez que lo contó, Jim dijo que habían cabalgado sobre él hasta Nueva Orleans, y después, cada vez que lo relataba iba exagerando la nota, hasta afirmar que le habían hecho correr alrededor del mundo, dejándole mortalmente cansado, y que tenía la espalda molida por la cabalgata. Jim se enorgullecía de la aventura. Los negros venían desde muy lejos para oírle contar a Jim lo ocurrido, y él fue mejor visto en el país que cualquier otro. Los negros forasteros se quedaban mirándole boquiabiertos como si vieran una maravilla. Los negros siempre hablan de brujas por la noche, junto al fuego de la cocina; pero, cuando uno hablaba de esas cosas y Jim estaba presente, exclamaba: «¡Hum! ¿Qué sabes tú de brujas?», y aquel negro se callaba cohibido y tenía que quedarse a un lado. Jim conservaba una moneda de cinco centavos colgada del cuello con un cordel y decía que era un amuleto que le dio el diablo con sus propias manos, diciéndole que podría curarlo todo con él y llamar a las brujas siempre que quisiera, pero Jim nunca explicó qué tenía que decir. Los negros acudían de todas partes para dar a Jim lo que tenían a cambio de la satisfacción de ver la moneda de cinco centavos, pero no la tocaban, porque había pasado por las manos del demonio. Jim casi dejó de servir para criado, porque tenía muchos humos desde que había visto al diablo y había llevado a las brujas.

Cuando Tom y yo llegamos a la cumbre de la colina, miramos hacia abajo, al pueblo, y vimos tres o cuatro luces que parpadeaban indicando tal vez las casas que tenían enfermos. Las estrellas, por encima de nosotros, lucían brillantes, y allá abajo, junto al pueblo, el río, de una milla de ancho, muy quieto e impresionante. Descendimos de la colina y encontramos a Joe Harper y Ben Rogers, y a dos o tres muchachos más, ocultos en la vieja tenería. De modo que desatracamos un esquife y remamos unas dos millas y media río abajo, hasta la enorme cicatriz en la ladera de la montaña, y bajamos a tierra.

Nos fuimos a unos arbustos y Tom hizo jurar a todos que guardarían el secreto, y luego les mostró un agujero en la montaña, en la parte más frondosa de los arbustos. Después encendimos las velas y entramos a gatas. Avanzamos unas doscientas yardas y entonces apareció la cueva. Tom buscó entre los pasadizos y al fin pasó agachado por debajo de una pared en la que no se veía un agujero. Pasamos por un lugar muy estrecho y entramos en una especie de habitación húmeda y fría, y allí nos detuvimos. Tom dijo:

—Ahora formaremos esta banda de ladrones que se llamará «Banda de Tom Sawyer». El que quiera entrar en la banda ha de prestar juramento y escribir su nombre con sangre.

Todos aceptaron. De modo que Tom sacó una hoja de papel, en la que estaba escrito el juramento y lo leyó. Cada muchacho juraba lealtad a la banda, no revelar sus secretos y que, si alguien le hacía algo a un chico de la banda, el chico que recibiera la orden de matar a esa persona y a su familia debería hacerlo, y no comer ni dormir hasta haberles dado muerte y haber hundido una cruz en su pecho, que era el emblema de la banda. Nadie que no fuera de la banda podría utilizar esa señal, y el que lo hiciera sería enjuiciado, a la segunda vez que la utilizara sería muerto. Y si algún miembro de la banda revelaba sus secretos, le cortarían el cuello, quemarían su cuerpo y esparcerían las cenizas al viento; su nombre ensangrentado quedaría borrado de la lista, y la banda jamás volvería a pronunciarlo, sino que le añadía una maldición y lo olvidaría para siempre.

Todos dijeron que era un juramento acertado y preguntaron a Tom si se lo había inventado. El dijo que en parte sí, pero que el resto procedía de libros de piratas y de salteadores de caminos, y que lo tenían todas las bandas de categoría.

Algunos pensaron que sería magnífico matar también a las familias de los chicos que revelaran los secretos. Tom dijo que era una buena idea, de modo que tomó el lápiz y lo escribió. Luego dijo Ben Rogers:

—Pero Huck Finn no tiene familia... ¿Qué haríamos con él?

—Bien, ¿acaso no tiene un padre? —replicó Tom Sawyer.

—Sí, lo tiene, pero no hay modo de encontrarlo hoy día. Solía dormir borracho con los cerdos en la tenería, pero no se le ve por aquí desde hace más de un año.

Lo discutieron y querían excluirme porque decían que cada chico debía tener una familia o alguien a quien poder matar, que si no, sería injusto para con los otros. Bueno, a nadie se le ocurría algo, y estaban todos pensativos, quietos. Me entraron ganas de llorar, pero de pronto tuve una idea y les ofrecí a la señorita Watson: podrían matarla a ella. Todos dijeron entonces:

—¡Oh, ella sirve, sirve! ¡Conformes, Huck puede ser de la banda!

Luego todos se pincharon el dedo con un alfiler para sacar sangre para la firma, y yo puse mi señal en el papel.

—Pero —dijo Ben Rogers—, ¿a qué clase de actividades se dedicará esta banda?

—Nada más que a robar y a asesinar —dijo Tom.

—Pero ¿qué robaremos? ¿Casas, ganado o...?

—¡Puaf! Robar ganado y cosas por el estilo no es robar —dijo Tom Sawyer—. Nosotros no somos ladrones. Eso no va con nosotros. Somos salteadores de caminos. Detendremos diligencias y carruajes en el camino, con antifaces, y mataremos a la gente y les quitaremos los relojes y el dinero.

—¿Siempre hemos de matar gente?

—¡Oh, pues claro! Es mejor. Algunas autoridades opinan lo contrario, pero la mayoría considera, mejor matarlos, exceptuando a los que traigamos a la cueva para tenerlos prisioneros hasta que paguen rescate por ellos.

—¿Qué es eso del rescate?

—No lo sé, pero es lo que hacen. Lo he leído en los libros y, por lo tanto, es lo que hemos de hacer, naturalmente.

—Pero ¿cómo lo haremos si no sabemos qué es?

—¡Diablo, debemos hacerlo! ¿No te he dicho que lo dicen los libros? ¿Quieres hacerlo de otra manera de como lo dicen los libros y enredar las cosas?

—¡No basta con decirlo, Tom Sawyer! ¿Cómo van a ser rescatados esos individuos, si no sabemos cómo se hace? ¡A eso quiero llegar yo! ¿Qué supones tú que es?

—Pues no lo sé; pero tal vez, si los retenemos hasta que sean rescatados, ello querrá decir que los retenemos hasta que estén muertos.

—Bien, eso ya me gusta más. Es la explicación. ¿Por qué no lo decías antes? Los tendremos hasta que mueran de rescate... Y menudo fastidio serán, comiéndoselo todo y tratando a cada momento de huir...

—¡Hablas mucho, Ben Rogers! ¿Cómo quieres que se escapen, si habrá un centinela vigilándolos, dispuesto a disparar en cuanto se pongan pesados?

—Un centinela... ¡Bien, bien! ¿De manera que alguien tendrá que pasarse la noche en blanco, simplemente para vigilarlos? ¡Me parece una tontería! ¿Por qué no puede uno coger un garrote y rescatarlos apenas lleguen aquí?

—Sencillamente, porque esto no lo dicen los libros, ¿entiendes? Y ahora, Ben Rogers, ¿quieres hacer las cosas como hay que hacerlas o no? ¿No comprendes que la gente que escribió los libros sabe lo que hay que hacer? ¿Te figuras que tú puedes enseñarles algo nuevo? ¡Ni lo sueñes! No, señor, los rescataremos siguiendo el método.

—Bueno, no me importa; pero de todos modos digo que es una tontería. Oye, ¿mataremos también a las mujeres?

—Mira, Ben Rogers, si yo fuese tan ignorante como tú, procuraría disimularlo. ¿Matar a las mujeres? No... Nadie ha visto cosa parecida en los libros. Se las lleva a la cueva y se las trata con cortesía, y poco a poco ellas se enamoran de ti y nunca quieren volver a sus casas.

—Pues, en este caso, conforme; pero no sé qué resultado dará. No tardaremos en tener la cueva tan atestada de mujeres y de individuos en espera de ser rescatados, que no nos quedará sitio para los ladrones. Pero adelante; no diré nada más.

El pequeño Tommy Barnes se había dormido y cuando le despertaron, estaba asustado y lloró, y dijo que quería ir a casa, con su mamá, y que ya no quería ser ladrón.

Todos se rieron de él y le llamaron *bebé* llorón, cosa que le puso furioso, y dijo que se iría derechito a contar los secretos. Pero Tom le dio cinco centavos para que se callara, diciendo que todos iríamos a casa y que nos reuniríamos a la semana siguiente para matar a alguien y robar a algunos.

Ben Rogers dijo que no podía salir mucho, sólo los domingos, de modo que quería empezar el siguiente domingo, pero los chicos dijeron que estaría mal hacerlo en domingo, y así quedó la cosa. Acordamos reunirnos y señalar un día en cuanto nos fuera posible, y después elegimos a Tom Sawyer como primer capitán y a Joe Harper como segundo capitán de la banda, y volvimos a casa.

Trepé por el cobertizo y entré por la ventana de mi cuarto poco antes de que rompiera el alba. Mis ropas nuevas estaban manchadas de grasa y barro, y yo estaba cansadísimo.

3

La señorita Watson me dio un buen rapapolvo por la mañana, a causa de un traje, pero la viuda no me regañó, sino, al contrario, lo limpió de grasa y barro; tenía un aspecto tan triste, que decidí portarme bien un rato, si podía. Luego la señorita Watson me llevó al gabinete y rezó, pero no consiguió nada. Me dijo que rezara cada día y que conseguiría lo que quisiera, pero no fue así. Lo probé. Una vez conseguí una caña de pescar, pero los anzuelos no. ¿De qué me serviría sin los anzuelos? Insistí en pedir los anzuelos tres o cuatro veces, pero no daba resultado. Al fin un día pedí a la señorita Watson que lo intentara por mi cuenta, pero dijo que yo era un estúpido. Nunca me dijo por qué, ni logré adivinarlo.

Una vez me senté en el bosque y estuve pensando en ello largo rato. Me dije que, si uno puede conseguirlo todo rezando, ¿por qué el diácono

Winn no recuperaba el dinero perdido con los cerdos?, ¿por qué la viuda no recuperaba la caja de rapé que le habían robado? No, me dije, no es cierto. Fui a contárselo a la viuda, y ella me dijo que lo único que se consigue rezando son «bienes espirituales». Esto era demasiado para mí, pero me explicó a qué se refería: yo debía ayudar a los demás y hacer por ellos lo que pudiera, cuidarlos siempre y no pensar nunca en mí mismo. Supuse que incluía a la señorita Watson.

Fui al bosque y reflexioné sobre ello largo rato, pero no encontré la ventaja, excepto para el prójimo; de modo que acabé por decidir no preocuparme más y olvidarlo.

A veces la viuda me llevaba aparte y me hablaba de la providencia de un modo que se le hacía a uno la boca agua, pero luego, al siguiente día, la señorita Watson venía y lo echaba todo a perder. Por lo que pude ver, había dos providencias: uno podía pasarlo en grande con la providencia de la viuda, pero, si le pillaba la de la señorita Watson, estaba perdido. Lo pensé y decidí que me quedaba con la de la viuda, si me quería, aunque no comprendí si a su providencia iba a convenirle contar conmigo, teniendo en cuenta que yo era tan ignorante, ordinario y despreciable.

A papá no lo habían visto desde hacía más de un año, y esto me resultaba cómodo. No quería volver a verlo. Solía vapulearme cuando estaba sobrio y podía echarme la mano encima, aunque yo tomaba la precaución de merodear casi siempre por el bosque cuando él estaba cerca. Bueno, pues por ese tiempo le encontraron ahogado en el río, a unas doce millas más abajo del pueblo, según dijo la gente. Por lo menos creyeron que era él; dijeron que el ahogado tenía su misma estatura, iba andrajoso y llevaba los cabellos muy largos —como papá—, pero no pudieron comprobar si era su rostro, porque había estado en el agua tanto tiempo, que ya no tenía aspecto de rostro. Dijeron que flotaba de espaldas en el agua. Le enterraron en la ribera. Pero mi tranquilidad fue corta, porque se me ocurrió algo. Sabía perfectamente que un ahogado no flota de espaldas, sino boca abajo. Entonces comprendí que no era papá, sino una mujer vestida con ropas de hombre. Nuevamente me inquieté. Pensé que el viejo reaparecería pronto, y eso no me hacía ninguna gracia.

Jugamos a bandidos durante un mes y luego dimití. Dimitieron todos. No habíamos robado a nadie, ni habíamos matado a ninguna persona, sólo lo fingíamos. Acostumbrábamos a salir de los bosques al trote y cargar sobre porqueros y las mujeres que llevaban hortalizas en carros al mercado, pero nunca les hacíamos daño. Tom Sawyer llamaba «lingote» a los cerdos y «alhajas» a los nabos, y llegábamos a la cueva y charloteábamos sobre lo que habíamos hecho y las numerosas personas que habíamos matado y señalado. Pero yo no vi provecho alguno en todo aquello.

Una vez Tom mandó a un chico al pueblo para que lo recorriera con un palo encendido, que él llamaba «grito de combate» (que era la señal para que la banda se reuniera), y luego dijo que tenía noticias secretas, que le habían dado sus espías, de que al día siguiente un grupo de mercaderes españoles y árabes ricos se disponían a acampar en la hondonada de la cueva, con doscientos elefantes, seiscientos camellos y más de mil acémilas cargadas de brillantes, y que sólo tenían una guardia de cuatrocientos soldados, de modo que les tenderíamos una emboscada, según lo llamaba Tom, y mataríamos a todos y nos llevaríamos el botín. Dijo que debíamos prepararnos limpiando las espadas y las armas. Nunca podía asaltar ni un carro de nabos sin obligarnos a sacar brillo a las espadas y a las armas de fuego, aunque sólo eran listones y mangos de escobas, y ya podía uno limpiarlos hasta pudrirse, que al terminar no valían más que un puñado de cenizas, como antes. No creí que pudiéramos atacar a tantos españoles y árabes, pero quería ver los elefantes y los camellos, de modo que al día siguiente, sábado, estuve presente en las emboscadas, y cuando recibimos la señal salimos de los bosques y bajamos la colina. Pero no había españoles, árabes, elefantes ni camellos. Solamente había una merienda de la escuela dominical, y de la clase de párvulos, por añadidura. Los desparramamos y perseguimos a los niños hasta la hondonada, pero no conseguimos más que algunos buñuelos y mermeladas, aunque Ben Rogers consiguió una muñeca pepona y Joe Harper un libro de himnos y un folleto. Luego nos persiguió el maestro y nos obligó a entregarlo todo y a largarnos en seguida. No vi brillantes por ningún lado, y así se lo dije a Tom Sawyer. El dijo que los había a montones, y árabes, y elefantes, y todo lo demás. Pregunté cómo no los había visto yo. Replicó que, si yo no fuera tan ignorante y hubiera leído un libro llamado *Don Quijote,* lo sabría sin necesidad de preguntarlo. Dijo que todo se hacía por encantamiento. Dijo que allí había centenares de soldados y elefantes y un tesoro, pero que teníamos enemigos, a los que llamó magos, que lo convirtieron todo en chiquillos de la escuela dominical con el único propósito de fastidiarnos. Dije que bueno, que entonces lo que teníamos que hacer era atacar a los magos. Pero Tom dijo que yo era un zoquete.

—¿No entiendes —dijo— que un mago podría llamar a muchos genios y te haría polvo antes de que pudieras decir esta boca es mía? Son altos como un árbol y su cuerpo tan grande como una iglesia.

—Bueno —dije yo—, supón que pedimos ayuda a los genios nosotros. ¿No podríamos atacarlos entonces?

—¿Y cómo los conseguirías?

—No lo sé... ¿Cómo los consiguen ellos?

—Pues frotan una lámpara vieja de hojalata o una sortija de hierro y entonces aparecen los genios, acompañados de truenos, relámpagos y humo, y hacen todo lo que se les pide. Para ellos no es nada arrancar de cuajo una torre y atizar con ella a un maestro de la escuela dominical... o a cualquier otro hombre.

—¿Quién les hace actuar de este modo?

—Pues la persona que frote la lámpara o la sortija. Son los esclavos de quien frote la lámpara o la sortija y tienen que hacer lo que se les ordena. Si les ordena edificar un palacio de cuarenta millas de largo, con diamantes, y llenarlo de goma de mascar o lo que quiera, y traer a la hija del emperador de China para que se case con él, tienen que hacerlo... y además antes de que salga el sol al día siguiente. Y, otra cosa: deben pasear el palacio por todo el país en la dirección que tú quieras, ¿lo comprendes?

—Bien —dije yo—, pues a mí me parece que son un hatajo de idiotas por no quedarse el palacio para sí mismos en lugar de desprenderse de él. Y, además, si yo fuera uno de ellos, antes me iría a las chimbambas, que dejarlo todo para correr en cuanto alguien frotase una lámpara vieja de hojalata.

—¡Hay que ver cómo hablas, Huck Finn! Tendrías que acudir cuando la frotase, lo quisieras o no.

—¿Cómo? ¿Y dices que yo sería alto como un árbol y grande como una iglesia? De acuerdo, vendría, pero te apuesto a que obligaría a ese hombre a subirse al árbol más alto que hubiera en el país.

—¡Corcho, es inútil hablar contigo, Huck Finn! Por lo visto, no sabes nada... ¡Eres un zoquete!

Lo pensé durante dos o tres días y decidí averiguar si había algo cierto en aquello. Conseguí una vieja lámpara de hojalata y una sortija de hierro, y me interné en los bosques y froté hasta que sudé como un indio injun, haciendo cálculos de edificar un palacio y venderlo, pero todo fue inútil: no vino ningún genio. Pensé que todo era una mentira más de Tom Sawyer. Supuse que él creía en los árabes y los elefantes, pero en cuanto a mí, opino de otro modo. Me pareció que aquello tenía todos los síntomas de una escuela dominical.

4

Pasaron tres o cuatro meses; estábamos ya en pleno invierno. Casi todo el tiempo estuve en la escuela, y ahora sabía deletrear, leer y escribir un poco, y recitaba la tabla de multiplicar hasta seis por siete igual a treinta y cinco, y no creo que llegue más allá aunque viva eternamente. No confío mucho en las matemáticas, de todos modos.

Al principio odié la escuela, pero poco a poco llegué a soportarla. Cuando me cansaba demasiado hacía novillos, y la zurra que me daban al día siguiente me sentaba bien y me animaba. De modo que cuanto más iba al colegio más fácil resultaba. Empezaba a acostumbrarme también a las cosas de la viuda y no se me hacía tan cuesta arriba vivir en su casa. Vivir en una casa y dormir en una cama me esclavizaba bastante, pero antes de que llegara el frío solía escabullirme para dormir en los bosques, y eso era un respiro para mí. Prefería mi vida anterior, pero, según iban las cosas, empezaba a gustarme un poquitín la nueva. La viuda decía que yo avanzaba despacio, pero seguro y con resultado satisfactorio. Dijo que no se avergonzaba de mí.

Una mañana, en el desayuno, tiré el salero. Cogí un pellizco para echarlo por encima de mi hombro izquierdo y contrarrestar así la mala suerte, pero la señorita Watson fue más rápida que yo y me lo impidió, diciendo: «¡Aparta las manos, Huckleberry... siempre estás haciendo destrozos!» La viuda intercedió por mí, pero eso no iba a librarme de la mala suerte, bien lo sabía yo. Salía después de desayunar preocupado y alterado, preguntándome qué iba a sucederme. Hay maneras de ahuyentar algunas clases de mala suerte, pero ésta era de otra clase. No traté de hacer nada, sino que continué andando abatido y alerta.

Bajé por el jardín de delante y me acerqué al fondo para atravesar la alta valla. Había una pulgada de nieve en el suelo y vi las pisadas de alguien. Alguien había venido desde la cantera y había rondado un rato para dar después la vuelta a la valla del jardín. Era curioso que no hubiera entrado después de estar allí tanto rato. No me lo explicaba. Sí, era muy extraño. Iba a seguir el rastro, pero me incliné para examinar antes las huellas. Al principio no noté nada, pero después sí: en el talón de la bota izquierda había la señal de una cruz hecha con clavos grandes, para ahuyentar al demonio.

Me incorporé y bajé disparado colina abajo. De vez en cuando miraba atrás por encima del hombro, pero no vi a nadie. Llegué a casa del juez Thatcher tan de prisa como me fue posible. El dijo:

—Muchacho, estás sin aliento. ¿Has venido a buscar el interés?

—No, señor —dije yo—; ¿tengo algo?

—Oh, sí: el que corresponde a medio año. Más de ciento cincuenta dólares. Una fortuna para ti. Será mejor que me dejes invertirlos junto con los otros seis mil, porque, si te los llevas, los gastarás.

—No, señor —contesté yo—. No quiero gastarlos. No los quiero... ni tampoco los seis mil dólares. Quiero que los acepte usted. Deseo darle los seis mil... todo.

Parecía sorprendido. No acababa de entenderlo. Dijo:

—Pero, ¿qué quieres decir, muchacho?

—Por favor, no me haga preguntas. Los aceptará, ¿no?

Contestó:

—Estoy intrigado, la verdad. ¿Ocurre algo?

—Acéptelos, por favor —dije yo—, y no me haga preguntas... Así no tendré que contarle mentiras.

Pensó unos momentos y luego exclamó:

—Aaah, me parece que ya lo comprendo. Quieres vender la propiedad, no dármela. Esta es la idea correcta.

Escribió algo en un papel, lo leyó y dijo:

—Mira... Dice: «por un precio». Eso significa que te lo he comprado pagándolo. Aquí tienes un dólar. Ahora fírmalo.

Firmé y me marché.

Jim, el negro de la señorita Watson, tenía una pelota de pelo tan grande como un puño, que había sacado del estómago de un buey, y él la usaba para magia. Decía que dentro había un espíritu que lo sabía todo. Por esto acudí aquella noche a verle y le dije que sabía que papá había vuelto porque encontré sus huellas en la nieve. Lo que yo quería saber era qué iba a hacer y si se quedaría. Jim sacó su pelota de pelo, pronunció algunas palabras sobre ella, luego la levantó y la dejó caer en el suelo. Parecía bastante sólida y rodó sólo una pulgada. Jim lo intentó de nuevo y luego otra vez, y la pelota hizo lo mismo. Jim se arrodilló, aplicó la oreja contra ella y escuchó. Pero fue en vano: dijo que la pelota se negaba a hablar. Explicó que a veces no hablaba sin dinero. Le dije que tenía una vieja moneda falsa de un cuarto de dólar que no servía porque se veía un poco el latón a través de la plata, y que, de todas maneras, no pasaría aunque no se viese el latón porque estaba tan gastada, que rezumaba grasa y se notaría que no era buena. Pensó que era mejor no hablar del dólar que me había dado el juez.

Dije que era dinero malo, pero que tal vez la pelota de pelo lo aceptara porque no notaría la diferencia. Jim olfateó la moneda, la mordió, la frotó y dijo que se las arreglaría para que la pelota de pelo creyera que era buena. Dijo que rajaría una patata irlandesa y metería dentro la moneda durante toda la noche, y que al día siguiente habría desaparecido la grasa y no se vería el latón, por lo que cualquiera del pueblo la aceptaría, y más aún una pelota de pelo. Bueno, yo sabía ya que una pelota podía conseguir esto, pero se me había olvidado.

Jim puso la moneda debajo de la pelota de pelo y se agachó para escuchar. Esta vez dijo que la pelota estaba conforme, y que si yo quería me diría la buenaventura. Dije que adelante. De modo que la pelota de pelo habló a Jim y éste me lo contó a mí:

—Tu padre no sabe todavía lo que va a hacer. A veces piensa en marcharse y a veces piensa en quedarse. Lo mejor es tomarlo con calma y dejar que el viejo decida. Hay dos ángeles que le guardan. Uno de ellos es blanco y resplandeciente, el otro es negro. El blanco le lleva por el camino recto un rato, pero el negro lo echa todo a perder en seguida. Nadie puede saber cuál de los dos se lo llevará por fin. Tú estás bien. Vas a tener muchas tribulaciones y considerables alegrías. A veces te harán daño, a veces enfermarás, pero siempre te pondrás bien. Pasarán dos chicas por tu vida. Una es rubia y la otra morena. Una es rica y la otra es pobre. Primero te casarás con la pobre y después con la rica. Debes estar lo alejado que puedas del agua y no corras peligros, porque está escrito que van a colgarte.

Aquella noche, cuando encendí mi vela y subí a mi cuarto, encontré allí a papá sentado... ¡El mismo!

5

Había cerrado la puerta. Me volví y allí estaba. Solía tenerle miedo, pues me pegaba mucho. Creo que también entonces estaba asustado, pero al minuto me di cuenta de que me había equivocado. Es decir, tuve un sobresalto, como quien dice, cuando la respiración se me cortó...

pues había aparecido de manera tan inesperada, pero en seguida vi que yo no estaba tan asustado como para preocuparme.

Tenía cincuenta años y los aparentaba. Llevaba los cabellos largos, enredados y sucios. A través de la pelambrera se veían sus ojos relucientes, como si estuvieran detrás de enredaderas. Su pelo era negro, no gris, y también negras eran sus patillas, largas y embrolladas. Su cara no tenía color, en los sitios donde se le veía la cara: era blanca, mas no del color blanco de cualquier otro hombre, sino de un blanco que ponía enfermo a uno, un color blanco que llenaba a uno de temblores. En cuanto a sus ropas, no eran más que andrajos. Apoyaba un tobillo sobre la otra rodilla; la bota de ese pie estaba reventada y le salían fuera dos dedos, que él movía de vez en cuando. Su sombrero estaba en el suelo; un viejo chambergo negro, de copa hundida, como si fuera una tapadera.

Me quedé de pie mirándolo. El seguía sentado observándome, la silla un poco echada hacia atrás. Dejé la vela. Me di cuenta de que la ventana estaba abierta. Deduje que había subido por el cobertizo. No dejaba de mirarme de pies a cabeza. Al poco rato dijo:

—¡Qué ropa tan almidonada! Te creerás un señorón, ¿verdad?

—Puede que lo sea y puede que no —contesté.

—A mí no me hables en ese tono —dijo él—. Se te han metido muchas tonterías en la cabeza mientras he estado fuera, pero ya te las quitaré, hasta que acabe contigo. Dicen que ahora estás educado, que sabes leer y escribir. Te crees mejor que tu padre, ¿verdad? Porque yo no sé, ¿eh? ¡Ya te ajustaré las cuentas! ¿Quién te dijo que podías enredarte en esas bobadas de lechuguino, eh? ¿Quién te lo digo?

—La viuda; ella me lo dijo.

—La viuda, ¿eh? ¿Y quién dijo a la viuda que podía meterse donde no le importaba?

—No se lo dijo nadie.

—¡Ya le enseñaré yo a ocuparse de sus asuntos! Y oye: tú dejas la escuela, ¿entiendes? Ya enseñaré yo a la gente a educar a un chico a que se dé aires de importancia para demostrar a su padre que vale más que él. ¡Verás si te encuentro rondando ese colegio! ¿Me oyes? Tu madre se murió sin saber leer ni escribir. Nadie de la familia sabía antes de morirse. Y yo tampoco sé; y ahora vienes tú pavoneándote. No lo aguanto, ¿lo oyes? Oye... A ver, que te oiga leer.

Cogí un libro y empecé a leer algo sobre George Washington y las guerras. Cuando llevaba cosa de medio minuto leyendo, él cogió el libro de un manotazo y lo arrojó al otro extremo del cuarto, diciendo:

—Así es: sabes leer. Lo dudé cuando me lo dijiste. Mira, deja ya de darte importancia. No lo consiento. Estaré vigilándote, bribón, y si te pesco cerca de ese colegio te daré una buena paliza. Antes de que te des cuenta te habrán hecho cura también. ¡Nunca he visto un hijo como tú!

Cogió una estampita azul y amarilla, con unas vacas y un niño, y dijo:

—¿Qué es eso?

—Algo que me dieron por saberme las lecciones.
El la rompió y dijo:
—Te daré algo mejor... te daré una paliza.
Estuvo allí sentado, gruñendo y murmurando durante un minuto, y luego dijo:
—¡Valiente pisaverde perfumado...! ¿Verdad? Cama, mantas y un espejo; con alfombra y todo... Y tu pobre padre tiene que dormir con los cerdos en la tenería. No he visto hijo como tú. Apuesto a que te quitaré algunas de esas costumbres blandengues antes de que acabe contigo. ¡Vaya, me pones enfermo con tus aires de suficiencia! Dicen que eres rico... ¿eh? ¿Cómo es eso?
—Pues mienten... Sí, mienten...
—Oye, tú, cuidado con hablarme así. Aguanto todo lo que puedo, de modo que no me provoques. En los dos días que llevo aquí, en el pueblo, sólo he oído hablar de tus riquezas. También lo decían río abajo. Por esto he venido. Mañana me darás ese dinero... Lo quiero.
—No tengo dinero.
—¡Mientes! Lo tiene el juez Thatcher. Se lo pides porque lo quiero.
—Te digo que no tengo dinero. Pregúntaselo al juez Thatcher; te dirá lo mismo.
—Bueno, se lo preguntaré y tendrá que escucharme o lo sentirá. Oye, ¿cuánto llevas en el bolsillo? Lo quiero.
—Solamente un dólar, y lo necesito para...
—No me importa para qué lo necesitas... ¡Anda, suéltalo!
Lo cogió y lo mordió para asegurarse de que era bueno. Luego dijo que bajaba al pueblo a beber whisky, que no había echado ningún trago en todo el día. Cuando saltó al cobertizo, asomó otra vez la cabeza y me maldijo por ser un presumido y querer ser mejor que él. Cuando calculé que ya se había ido, volvió a aparecer su cabeza y me recordó lo del colegio, porque, si iba, me daría una paliza fenomenal.
Al día siguiente, estando borracho, fue a casa del juez Thatcher y le insultó tratando de sacarle el dinero, pero no pudo, y entonces juró que le obligaría legalmente.
El juez y la viuda apelaron a la ley para conseguir de los tribunales que me apartaran de él y que uno de ellos fuera mi tutor, pero había llegado un nuevo juez que no conocía a mi viejo, de manera que dijo que los tribunales no debían intervenir para separar a las familias si podían evitarlo, y afirmó que él no haría nada para apartar a un hijo de su padre. Así que el juez Thatcher y la viuda tuvieron que dejar el asunto.
Eso hizo enormemente feliz al viejo. Dijo que me azotaría hasta que me pusiera morado si no reunía el dinero para él. Pedí prestados tres dólares al juez Thatcher y papá los cogió para emborracharse, y no paró de ir de un lado a otro, maldiciendo y gritando. Estuvo así hasta la medianoche, recorriendo el pueblo con una cazuela de hojalata; luego lo metieron en la cárcel, y al día siguiente lo llevaron ante el juez y volvieron a encerrarlo durante una semana. Pero mi padre dijo que estaba satisfecho, que era el dueño de su hijo y que él se las pagaría.

Cuando salió, el nuevo juez dijo que iba a hacer de él un hombre. De modo que le llevó a su casa, le dio ropa limpia y lo invitó a desayunar, a comer y a cenar con su familia, y, como quien dice, le trató como si fuera su hermano. Después de cenar le habló de la sobriedad y otras cosas, hasta que el viejo se echó a llorar diciendo que había sido un loco y destrozado su vida, pero que iba a cambiar para ser un hombre del que nadie se avergonzaría y que esperaba que el juez le ayudara y no le mirase con desprecio. El juez dijo que sentía deseos de abrazarlo por esas palabras, de modo que también lloró, así como su esposa. Papá dijo que nadie le había comprendido nunca, y el juez afirmó que así lo creía. El viejo dijo que un hombre caído necesitaba simpatía, y el juez aseguró que así era en efecto, de manera que volvieron a llorar. Y cuando fue hora de acostarse el viejo se levantó y dijo, alargando la mano:

—Mírenla, caballeros y señoras todos. Estréchenla. Esa es la mano que fue de un cerdo, pero ya no lo es. Es la mano de un hombre que ha empezado una nueva vida y que morirá antes de echarse atrás. Recuerden estas palabras... No olviden que las he dicho. Ahora es una mano limpia; estréchenla, no tengan miedo.

De modo que se la estrecharon uno tras otro y lloraron un poquito más. La esposa del juez la besó. Luego el viejo firmó un juramento... poniendo su marca. El juez dijo que era el momento más sagrado que recordaba, o algo por el estilo. Luego metieron al viejo en una habitación muy bonita, la de los huéspedes, y por la noche mi padre sintió una sed horrible, por lo que se descolgó al tejado del porche y se rompió el brazo izquierdo por dos sitios, y estaba casi helado de frío cuando lo encontraron, después de salir el sol. Y, cuando subieron a echar una mirada a la habitación de los huéspedes, tuvieron que pensarlo para entrar en ella.

El juez se sintió amargado. Dijo que suponía que se podría reformar al viejo con un rifle, pero que él ignoraba otro sistema.

6

El viejo no tardó en levantarse y andar por todas partes y entonces llevó al juez Thatcher ante los tribunales para obligarle a entregar aquel dinero, y me ajustó las cuentas también por no dejar de ir al colegio. Me atrapó un par de veces y me vapuleó, pero yo seguí yendo a la escuela y casi siempre pude darle esquinazo. No es que me gustara mucho ir al colegio, pero creo que entonces iba por fastidiar a papá. El juicio ante el tribunal era lento. Parecía que no iban a empezar nunca, de modo que yo de vez en cuando pedía prestados dos o tres dólares al juez para mi padre, para evitarme una paliza. Cada vez que tenía dinero se emborrachaba y, cuando se emborrachaba, armaba las de Caín en el pueblo, y cada vez que armaba las de Caín le encerraban en la cárcel. Estaba hecho para esa clase de cosas.

Empezó a rondar demasiado la casa de la viuda, y así se lo dijo ella al fin, y que si no dejaba de molestarla le buscaría complicaciones. Bueno, ¡cómo se puso él! Dijo que veríamos quién mandaba sobre Huck Finn. De modo que me esperó un día de primavera, me llevó río arriba a unas tres millas, en un esquife, y pasamos a la orilla de Illinois, donde había bosques y ninguna casa, excepto una vieja cabaña en un lugar donde era tan densa la arboleda, que uno no podía encontrarla si no sabía con precisión dónde estaba.

Le tenía siempre al lado y no tuve oportunidad de escaparme. Vivíamos en esa vieja cabaña y él cerraba la puerta con llave, que se guardaba debajo de la cabeza por la noche. Tenía un arma, robada, supongo, y pescamos y cazamos, y así vivíamos. De vez en cuando me encerraba y bajaba al almacén, a tres millas del embarcadero del vapor, y canjeaba caza y pesca por whisky, que traía a casa, donde se emborrachaba, y después me daba una zurra. La viuda no tardó en averiguar dónde estaba yo y mandó a un hombre para que intentara sacarme de allí, pero papá le ahuyentó con su escopeta y al poco tiempo me acostumbré a estar allí y me gustaba todo, menos las palizas.

Era cómodo gandulear todo el día, fumando y pescando, sin libros de estudio. Pasaron dos meses o más, mis ropas se convirtieron en sucios andrajos y me pregunto cómo pudo gustarme estar en la casa de la viuda, donde tenía que lavarme y comer en un plato, peinarme, acostarme y levantarme a la misma hora siempre, y estar siempre devanándome los sesos con un libro y tener a la señorita Watson molestándome continuamente. No quería volver más allí. Había dejado de jurar porque a la viuda le desagradaba, pero ahora volví a jurar porque papá no se opuso. En conjunto resultaba divertido pasar el tiempo en los bosques.

Pero al fin papá empezó a tomarle demasiado gusto al garrote y no pude aguantarlo. Iba lleno de cardenales. Se hicieron más frecuentes sus ausencias mientras me dejaba encerrado con llave. Una vez me dejó encerrado tres días. Aquella soledad era terrible. Pensé que se había ahogado y que nunca podría salir de la cabaña. Tuve miedo. Decidí buscar un modo de escapar. Había intentado salir de la cabaña otras veces, pero nunca lo conseguí. Ninguna ventana era lo bastante grande para que pudiera pasar por ella un perro. No podía salir por la chimenea porque era demasiado estrecha. La puerta era de roble macizo. Papá estaba atento a no dejar ningún cuchillo o cualquier cosa en la cabaña cuando se iba. Creo que registré el sitio más de cien veces. Bueno, lo hacía casi siempre porque era el único modo de pasar el tiempo. Pero al fin encontré algo: una sierra vieja y oxidada, sin mango. La encontré entre una viga y las chillas del techo. La engrasé y empecé a trabajar. Había una vieja manta de montar a caballo clavada en los troncos del fondo de la cabaña, detrás de la mesa, para evitar que pasara el viento por las rendijas y apagara la vela. Me metí debajo de la mesa, alcé la manta y empecé a serrar una parte del enorme tronco inferior, lo bastante ancho para que pudiera salir yo. Fue una tarea muy larga, pero ya iba llegando al fin cuando oí la escopeta de caza de papá en los bosques. Borré las huellas de mi trabajo, dejé caer la manta y oculté la sierra. A poco entró él.

No estaba de buen talante... o sea como de costumbre. Dijo que había estado en el pueblo y que todo le salía mal. Su abogado creía que ganaría el pleito y obtendría el dinero si es que algún día empezaba el juicio, pero que había muchas maneras de dar largas al asunto y que el juez Thatcher sabía cómo hacerlo. Y dijo que la gente creía que habría otro juicio para separarme de él y entregarme a la viuda como tutora, y que todos suponían que ella ganaría esta vez. Me impresioné bastante porque no quería volver a casa de la viuda para sentirme tan sujeto y civilizado, como ellos lo llamaban. Entonces el viejo empezó a maldecir todas las cosas y a todos aquéllos que le vinieron en mientes, y volvió a maldecirlos para asegurarse de que no se olvidaba de ninguno, y después puso el broche final con una especie de maldición general, incluyendo a un considerable número de personas de quienes ni sabía los nombres, llamándoles esto y aquello, y luego siguió maldiciendo.

Dijo que le gustaría ver cómo me llevaba la viuda. Dijo que vigilaría y que, si trataban de hacerle esta jugarreta, un lugar a seis o siete millas, donde esconderme, y que allí ya podrían buscarme, que tendrían que darse por vencidos. Esto volvió a intranquilizarme bastante, aunque sólo durante un minuto, porque pensé que no me tendría a su alcance cuando buscara esa oportunidad.

El viejo me obligó a ir al esquife para traer las cosas, que consistían en un saco de cincuenta libras de harina de maíz, un lomo de cerdo, municiones y un barril de cuatro galones de whisky, un libro viejo y dos periódicos para relleno, aparte de alguna estopa. Llevé una carga, volví y me senté en la proa del esquife a descansar. Lo pensé bien y decidí llevarme la escopeta y algunas cañas de pescar cuando huyera para esconderme en los bosques. Supuse que no me quedaría en un sitio, sino que atravesaría el país, casi siempre de noche, y me alimentaría con la pesca y la caza, para ir tan lejos, que ni el viejo ni la viuda pudieran volver a encontrarme.

Acabaría de serrar el agujero y me fugaría aquella noche si papá se emborrachaba lo suficiente, y pensé que sí lo haría. Estaba tan ensimismado, que olvidé el rato que llevaba allí quieto, hasta que el viejo aulló preguntando si me había dormido o ahogado.

Subí todas las cosas a la cabaña y ya era casi de noche. Mientras yo hacía la cena, el viejo echó un par de tragos, se acaloró y empezó de nuevo a blasfemar. Había bebido en el pueblo, estuvo toda la noche en el arroyo y ofrecía un aspecto desastroso. Cualquiera hubiera creído que era Adán, tan cubierto iba de barro. Cuando empezaba a hacerle efecto el alcohol, solía emprenderlas con el gobierno. Esta vez dijo:

—¡Y lo llaman gobierno! No hay más que mirarlo para ver lo que es... Ahí tenemos a la ley dispuesta a quitar a un hombre a su hijo... su propio hijo, que le costó tantos esfuerzos, ansiedades y gastos para criarlo. Sí, y cuando ese hombre ha criado a fondo a su hijo y le tiene a punto para trabajar y hacer algo para él, para que sea su descanso, la ley se lo quita. ¡Y a eso llaman gobierno! Pero no para ahí la cosa... La ley ampara a ese viejo juez Thatcher, le ayuda a quitarme lo que es de mi propiedad. Eso es lo que hace la ley. La ley coge a un hombre que vale seis mil dólares, y más, y le encierra en una trampa como esta cabaña, y le deja ir por ahí con ropas que no son dignas ni de un cerdo. ¡Y a eso llaman gobierno! Con un gobierno así un hombre no puede hacer valer sus derechos. A veces me entran ganas de largarme del país para siempre. Sí, y se lo dije a ellos, se lo dije a la cara al viejo Thatcher. Me oyeron montones de gente que pueden atestiguarlo. Dije que por menos de dos centavos dejaba este maldito país y no volvía nunca. Esas fueron exactamente mis palabras. Dije que mirasen mi sombrero, si a eso puede llamársele sombrero, porque la tapa se levanta y el resto me cae hasta por debajo de la barbilla y, por lo tanto, no es sombrero ni mucho menos, sino parece como si mi cabeza saliere por la boca de una estufa. Miradlo, les dije, es el sombrero que llevo yo... uno de los hombres más ricos de este pueblo, si pudiera hacer valer mis derechos.

¡Ah, sí, sí; es un gobierno maravilloso, maravilloso...! Oye, mira, había un negro libre allá, en Ohio, un mulato casi tan blanco como un blanco. Llevaba puesta la camisa más blanca que has visto y el sombrero más reluciente; y en ese pueblo todos los hombres tienen nuevos trajes, y llevaba un reloj de oro con cadena, un bastón de empuñadura de plata... Era el más espantoso nabab de cabellos grises de todo el Estado. Y... ¿lo adivinas? Dijeron que era profesor en una universidad y que hablaba toda clase de idiomas y lo sabía todo. Y no es eso lo peor. Dijeron que él podía votar en su ciudad. Bueno, eso me sacó de mis casillas. Pienso adónde irá a parar el país. Era día de elecciones y yo me disponía a ir a votar si no estaba demasiado bebido para llegar, pero cuando me dijeron que en este país hay un Estado en el que dejan votar a ese negro, me eché atrás. Dije que nunca volvería a votar. Esas fueron mis palabras; todos me oyeron. Y por mí el país puede pudrirse... Jamás volveré a votar mientras viva. ¿Y la desfachatez de aquel negro? ¡Pero si ni me habría dejado pasar, de no apartarle yo de mi camino! Dije a la gente que por qué no subastaban a aquel negro y lo vendían. Esto es lo que yo quiero saber. ¿Y supones qué contestaron? Pues dijeron que no podían venderlo hasta que llevase seis meses en el Estado, y que todavía no los llevaba. Ya ves... Eso es un ejemplo. Llaman gobierno al que no puede vender a un negro libre hasta que lleva seis meses en el Estado. Un gobierno que se llama a sí mismo gobierno, se comporta como un gobierno y se cree un gobierno y, sin embargo, tiene que quedarse de brazos cruzados durante seis meses enteros antes de apoderarse de un ladrón infernal, de un negro libre, que lleva camisa blanca y...

Papá siguió hablando en ese tono sin darse cuenta siquiera de adónde le llevaban sus viejas piernas, de modo que se cayó de cabeza por encima del barril de cerdo salado y se lastimó las dos espinillas, y el resto del discurso fue hecho de la manera más acalorada y fuerte respecto al negro y al gobierno, aunque también intercaló algunas maldiciones contra el barril. Dio muchos saltos por la cabaña, primero sobre una pierna y luego sobre la otra, cogiéndose una espinilla y luego la otra, y al fin descargó de pronto un formidable puntapié con el pie izquierdo contra el barril. Pero no estuvo acertado en ello, porque era la bota de la que salían dos dedos, de modo que lanzó un alarido que ponía los pelos de punta antes de caer al suelo y rodar por el polvo agarrándose los dedos de los pies. Las maldiciones de entonces dejaron chiquititas las anteriores. El mismo lo dijo más tarde. Había oído a Sowberry Hagan en sus mejores días, y dijo que le superaba, pero me parece que se pasó de la raya en sus jactancias.

Después de cenar, papá cogió el barril y dijo que había whisky suficiente para dos borracheras y un delirium tremens. Es lo que siempre decía. Supuse que antes de una hora estaría completamente borracho. Entonces le robaría la llave o saldría por el agujero que serraba en los troncos. Bebió y bebió, y al poco rato cayó desplomado sobre las mantas, pero la suerte no me favoreció. No se durmió por completo, sino que permaneció inquieto. Gimió, gruñó y se agitó de un lado a otro durante largo tiempo. Al fin no pude continuar manteniendo abiertos los ojos, y antes de saber qué me pasaba, caí profundamente dormido, con la vela encendida.

No sé cuánto rato estuve durmiendo, pero de repente sonó un grito horrible y me levanté. Allí estaba papá, enloquecido, saltando de un lado a otro, gritando algo acerca de serpientes. Dijo que se le subían por las piernas; entonces daba un salto y gritaba, y decía que una le había mordido en la mejilla... Pero yo no vi ninguna serpiente. Empezó a correr por la cabaña dando vueltas y aullando:

—¡Llévatela! ¡Llévatela, me está mordiendo en el cuello!

Jamás vi una mirada tan espantosa en los ojos de un hombre. No tardó en quedar agotado y cayó jadeante; luego rodó con sorprendente rapidez, lanzando puntapiés a diestro y siniestro, dando manotazos en el aire, gritando y diciendo que los demonios se apoderaban de él. Quedó extenuado al poco rato y estuvo quieto, gimiendo. Luego se quedó más quieto aún y no hizo sonido alguno. Pude oír a los búhos y a los lobos, a lo lejos, en los bosques, y la quietud era horripilante. El yacía en el rincón. Poco a poco se incorporó y escuchó con la cabeza ladeada. Dijo en voz muy baja:

—Plaf... plaf... plaf... ¡Las pisadas de la muerte! Plaf... plaf... plaf... Viene por mí, pero no quiero ir... ¡Oh, están frías...! ¡Suéltame...! ¡Deja en paz a ese pobre diablo!

Luego se puso a cuatro patas, se arrastró suplicándole que le dejara tranquilo, en enrolló en su manta y se escabulló debajo de la mesa de pino, suplicando aún, y entonces empezó a llorar. Le oía a través de la manta.

Poco después salió rodando y se puso en pie de un salto, con aire de loco, me vio y vino hacia mí. Me persiguió por la cabaña con una navaja, llamándome ángel de la muerte, diciendo que me mataría para que no volviera por él. Le supliqué diciéndole que yo era solamente

Huck, pero soltó una carcajada espeluznante, rugió y continuó la persecución. Una vez, cuando di media vuelta y le esquivé pasando por debajo de su brazo, me agarró por la espalda de la chaqueta y pensé que estaba perdido, pero me escabullí de la chaqueta tan rápido como el rayo y pude ponerme a salvo. Al fin le rindió el cansancio, se dejó caer al suelo, de espaldas contra la pared, y dijo que descansaría un minuto antes de matarme. Se sentó encima del cuchillo y dijo que dormiría para recobrar fuerzas, y que entonces ya veríamos quién era quién.

No tardó en dormirse. Yo cogí la vieja silla de asiento resquebrajado, me subí a ella con tanta cautela como pude para no hacer ruido y descolgué la escopeta. Introduje la baqueta para cerciorarme de que estaba cargada, luego la puse encima del barril de nabos, apuntando hacia papá, y me senté detrás en espera de que se despertara. ¡Qué lento y monótono se hizo el tiempo de la espera!

7

—¡Levántate! ¿Qué te propones?

Abrí los ojos y miré alrededor, tratando de averiguar dónde me encontraba. El sol había salido ya, y yo había estado profundamente dormido hasta entonces. Papá estaba de pie a mi lado, con aire torvo y aspecto enfermizo. Dijo:

—¿Qué haces con esa escopeta?

Pensé que no se acordaba de lo que él había estado haciendo, de manera que contesté:

—Alguien intentaba entrar y me preparé a recibirlo.

—¿Por qué no me despertaste?

—Lo intenté, pero fue inútil. No pude ni moverte.

—Bueno, bueno. No te quedes ahí parado charlando todo el día. Sal a ver si ha picado el anzuelo algún pez para desayunar. Iré dentro de un minuto.

Abrió la puerta y yo me dirigí a la orilla del río. Vi algunas ramas y cosas parecidas flotando corriente abajo, así como una corteza, de modo que calculé que el río iba a crecer. Pensé que me hubiera divertido de lo lindo si hubiera estado en el pueblo. La crecida de junio siempre me daba suerte, porque, en cuanto empieza, la corriente arrastra leña, cuerdas y trozos de troncos flotantes..., a veces hasta media docena de troncos, y entonces no hay más que cogerlos y venderlos a los almacenes madereros y al serradero.

Remonté la orilla vigilando a papá con un ojo y con el otro viendo qué podía pescar del agua. Entonces apareció una canoa. Era una preciosidad, de unos trece o catorce pies de larga, que flotaba alta como un pato. Me zambullí de cabeza desde la orilla, como una rana, vestido y todo, y nadé hacia la canoa. Esperaba que dentro hubiera alguien tendido, porque esto es lo que suele hacer la gente para engañar a los demás y, cuando uno se ha ilusionado con la canoa, asoman la cabeza y se ríen de uno. Pero no ocurrió esta vez. No había duda de que la embarcación iba a la deriva, de modo que me metí dentro y remé hacia la orilla. Pensé que el viejo se alegraría al verla..., pues bien valía diez dólares. Pero, cuando desembarqué, papá no estaba a la vista aún, y mientras metía la canoa en una maleta parecida a una garganta, casi cubierta por enredaderas y sauces llorones, se me ocurrió otra idea. Decidí esconderla para, en vez de dirigirme a los bosques cuando huyera, ir río abajo a unas cincuenta millas de distancia de la cabaña y acampar definitivamente en un sitio, para no pasarlo mal andando a pie.

Estaba bastante cerca de la cabaña y estuve asustado temiendo a cada momento que se acercara el viejo, pero tuve tiempo de esconderla. Luego salí y me asomé a ver desde detrás de unos sauces. Allá abajo estaba el viejo, en el camino, apuntando a un pájaro con su escopeta. De modo que no había visto nada.

Cuando se acercó, yo estaba atareado con la caña de pescar. Me regañó un poco por mi lentitud, pero le dije que me había caído al río y por eso había tardado tanto. Sabía que se fijaría en mis ropas mojadas y empezaría a hacer preguntas. Capturamos cinco peces y volvimos a casa.

Mientras estábamos echados, después de desayunar, para dormir la siesta, rendidos tanto él como yo, empecé a cavilar cómo evitar que papá y la viuda intentaran seguirme. Más seguro que confiar en la suerte sería alejarme lo bastante antes de que notaran mi ausencia. Pueden ocurrir muchas cosas. El caso es que al principio no vi la solución, pero al poco rato papá se incorporó para beber otro cazo de agua y dijo:

—La próxima vez que se acerque un hombre a husmear por aquí tienes que despertarme, ¿entiendes? Ese no venía a nada bueno. Le hubiera pegado un tiro. La próxima vez me despiertas, ¿comprendido?

Luego se tendió y cayó dormido..., pero lo que había dicho me dio la idea que necesitaba. Me dije que ya sabía cómo arreglármelas para que nadie me siguiera.

Alrededor de las doce anduvimos ribera arriba. La corriente era bastante rápida y arrastraba montones de troncos a la deriva debido a la crecida. A poco apareció parte de una balsa de troncos..., unos nueve sujetos entre sí. Salimos con el esquife y la remolcamos a tierra. Luego comimos. Otro que no hubiera sido papá habría pasado el resto del día recogiendo más madera, pero eso no iba con su estilo. Nueve troncos le bastaban para una vez. Quería ir corriendo a venderlos al pueblo. De modo que me encerró y se marchó en el esquife remolcando la balsa a las tres y media aproximadamente.

Calculé que esa noche no regresaría. Esperé hasta que supuse que estaba ya lejos y entonces me dediqué a aserrar el tronco otra vez. Antes de que él hubiera alcanzado la orilla opuesta del río, yo había salido por el agujero. Papá y su balsa eran solamente unas motas en el agua, a lo lejos.

Cogí el saco de harina de maíz y lo llevé al escondite de la canoa. Aparté las enredaderas y las ramas para meterlo dentro de la embarcación. Luego hice lo mismo con el lomo de cerdo y con el barril de whisky; me llevé todo el café y el azúcar que había, así como las municiones. También me llevé estopa, el cubo y la calabaza vinatera, un cazo y una taza de hojalata, mi vieja sierra y dos mantas, lo mismo que la cazoleta y la cafetera. Cogí las cañas de pescar, cerillas y otras cosas... Todo lo que valía un centavo. Dejé la cabaña limpia. Necesitaba un hacha, pero no había más que la del montón de leña, y yo sabía por qué la dejaba. Cogí por último la escopeta y así terminé con mi trabajo.

Había desgastado bastante el suelo al salir a rastras del agujero sacando tantas cosas. De modo que lo arreglé del mejor modo posible desde fuera removiendo la tierra para borrar las huellas y el serrín. Luego coloqué el tronco en su sitio y coloqué debajo dos piezas una contra otra para apuntalarlo, porque en aquella parte formaba curva y apenas tocaba el suelo. Desde cuatro o cinco pies de distancia, de no saberse que había sido serrado, ni siquiera se notaba. Además, estaba en la parte de atrás de la cabaña y era improbable que alguien se acercara por allí.

Había hierba en el camino hasta el sitio donde estaba la canoa, de modo que no dejé huellas. Me di una vuelta para comprobarlo. Llegué a la orilla y miré hacia el río. Todo estaba tranquilo. Cogí la escopeta y me interné en los bosques, donde cacé algunos pájaros. Entonces vi un cerdo salvaje. En cuanto los cerdos se escapan de las granjas de la pradera, no tardan en volverse salvajes. Le pegué un tiro y me lo llevé al campamento.

Cogí el hacha y destrocé la puerta. Para conseguirlo tuve que descargar bastantes golpes sobre ella. Cogí el cerdo y lo llevé junto a la mesa, donde le corté el cuello con el hacha y lo dejé desangrándose en tierra. Digo tierra porque era tierra, compacta y sin tablas. Después cogí un saco viejo, lo llené de piedras grandes, todas las que pude arrastrar, y, empezando desde donde estaba el cerdo, lo arrastré hasta la puerta y a través de los bosques hasta el río, donde lo eché y estuve viendo cómo se hundía. Deseé que Tom Sawyer estuviera allí, pues sabía que a él le interesaría ese asunto, y le habría dado algunos toques de adorno con su fantasía. Nadie podría despacharse a gusto en esto tan bien como lo haría Tom Sawyer.

Finalmente me arranqué algunos cabellos, ensangrenté el hacha y la colgué en el rincón. Luego cogí el cerdo, lo apreté contra mi chaqueta (para que no goteara), hasta que estuve bastante más abajo de la cabaña, y entonces lo arrojé al río. Se me ocurrió otra cosa, de manera que fui a la canoa en busca del saco de comida y la sierra, y los llevé a la cabaña. Puse el saco donde solía estar antes y lo agujereé por la parte inferior con la sierra, porque no había cuchillos ni tenedores allí, ya que papá utilizaba para todo su navaja. Luego transporté el saco unas cien yardas por la hierba y a través de los sauces, al este de la casa, hacia un lago de aguas poco profundas, que tenía cinco millas de ancho y estaba lleno de juncos... y de patos también, dicho sea de paso, en la temporada. Había, al otro lado, un cenegal o caleta que se extendía hasta varias millas de distancia, no sé dónde, pero en dirección contraria al río. La harina de maíz caía del saco y dejaba un rastro hasta el lago. Allí dejé también la amoladera de papá, para darle apariencias de accidente. Luego tapé el agujero del saco atándolo con un cordel, para que dejara de soltar carga y me lo llevé, junto con la sierra, a la canoa.

Empezaba a anochecer, de modo que dejé correr la canoa río abajo, junto a unos sauces llorones que caían sobre la orilla, y esperé a que saliera la luna. Entonces comí un bocado y me eché en la canoa para fumar una pipa y hacer proyectos. Me dije que seguirían el rastro del saco lleno de piedras hasta la orilla y que entonces dragarían el río buscándome. Y que irían tras el reguero de harina hasta el lago y recorrerían la caleta que había al otro lado en busca de los ladrones que me mataron y se llevaron las cosas. En el río no buscarían más que mi cadáver. Pronto se darían por vencidos y dejarían de pensar en mí.

Podía detenerme donde quisiera. La isla Jackson me gusta. La conozco bastante bien, y allí nunca va nadie. Además, desde ella podría ir en barca al pueblo por las noches para merodear y coger las cosas que necesitara. La isla Jackson era el sitio adecuado.

Me sentí bastante cansado y sin darme cuenta me dormí. Al despertar, por un momento no supe dónde estaba. Me incorporé. El río parecía anchísimo. La luna brillaba tanto, que hubiera podido contar los troncos a la deriva que descendían flotando, negros, quietos, a centenares de yardas de la orilla. Todo estaba mortalmente tranquilo, parecía ser tarde y «olía» a tarde. Ya saben que quiero decir... No conozco las palabras para hacerme entender.

Bostecé a gusto, me estiré y ya iba a desamarrar la canoa para marcharme, cuando oí un ruido distante al otro lado del agua. Escuché. Pronto lo reconocí. Era el ruido opaco y monótono que producen los remos funcionando en las chumaceras cuando la noche está quieta. Atisbé por entre las ramas de los sauces y... allí estaba. Era un esquife al otro lado del agua. No podía saber cuántos iban dentro. Seguía avanzando y, cuando estuvo delante de mí, vi que solamente iba en él un hombre. Pensé que acaso fuera papá, pero no le esperaba. Pasó cerca, llevado por la corriente, y se aproximó a la orilla de agua tranquila y pasó tan cerca de mí, que hubiese podido tocarle con la escopeta. Pues sí era papá... y, además, sobrio, a juzgar por la manera cómo manejaba los remos.

No perdí tiempo. Al minuto siguiente me dejaba deslizar corriente abajo con suavidad, pero rápido, al amparo de la umbrosa orilla. Recorrí unas dos millas y media y luego giré un cuarto de milla o más hacia el centro del río, porque no tardaría en pasar el embarcadero y la gente podría verme y llamarme. Me metí entre los maderos que iban a la deriva y después me tendí en el fondo de la canoa y dejé que flotase.

Estuve descansando largo rato, fumando en pipa y contemplando el cielo, que aparecía limpio de nubes. El cielo parece muy profundo cuando uno está echado de cara hacia la luna. Jamás lo había observado. ¡Y desde qué lejos se oye en esas noches sobre el agua! Oí a gente que hablaba en el embarcadero del vapor. Oí lo que decían, hasta la última palabra. Un hombre afirmaba que se aproximaban los días largos y las noches cortas. Otro dijo que se figuraba que aquélla no era precisamente una de las noches cortas... y rieron, él lo repitió y volvieron a reírse. Después despertaron a otro, se lo contaron y rieron, pero él no. Les respondió bruscamente y dijo que le dejaran en paz. El primer hombre que había hablado declaró que pensaba contárselo a su mujer, lo hizo y a ella le pareció muy chistoso, aunque añadió que no era nada comparado con las cosas tan graciosas que decía en sus años mozos. Oí decir a un hombre que eran casi las tres y que esperaba que la claridad del día no tardara en llegar más de una semana. Después la conversación fue quedando atrás y ya no pude entender las palabras, aunque sí percibía el murmullo de sus voces, y de vez en cuando una carcajada distante.

Estaba más abajo del embarcadero, a unas dos millas y media río abajo, cubierta de arboleda y levantándose en el centro del río, grande, sombría, sólida, como un vapor sin luces. No se veían indicios del banco, que ahora estaba lejos del agua.

No me costó mucho llegar allí. Pasé el cabo a gran velocidad, por lo rápido de la corriente, y luego entré en agua mansa y desembarqué en el lado orientado hacia la orilla de Illinois. Metí la canoa en una hendidura profunda de la orilla que yo conocía. Tuve que apartar las ramas de los sauces para hacerla entrar y, cuando la amarré, desde fuera nadie habría podido ver la canoa.

Subí a sentarme sobre un tronco en el cabo de la isla y miré el gran río y los troncos negros a la deriva, y luego hacia el pueblo, a tres millas de distancia, donde parpadeaban tres o cuatro luces. A una milla río arriba había una enorme y mostruosa balsa que descendía con una linterna en el centro. La vi acercarse y, cuando estaba casi frente adónde me encontraba yo, oí decir a un hombre: «¡Vamos, remos de popa! ¡Virad hacia estribor!»

Lo oí tan claramente como si el hombre estuviera a mi lado.

Había un ligero color gris en el cielo, de modo que me interné en los bosques y me eché en el suelo para dormir un rato antes de desayunar.

8

Cuando desperté, el sol estaba tan alto, que calculé que eran más de las ocho. Estuve tumbado en la hierba, bajo la sombra, pensando en cosas y sintiéndome descansando, feliz y contento. Podía ver el sol por un par de rendijas, pero en su mayoría eran árboles enormes los que había en derredor. En tierra había sitios moteados, allí donde la luz quedaba matizada por las hojas, demostrando la ligera brisa que corría más arriba. Un par de ardillas posadas en una rama charloteaban amistosamente.

Me sentí perezoso y a gusto... No quería levantarme para preparar el desayuno... No quería levantarme para preparar el desayuno. Bueno, ya estaba adormilándome otra vez cuando me pareció oír el profundo sonido de un ¡booom! río arriba. Me incorporé, apoyado en un codo, y escuché; pronto volví a oírlo. Me puse en pie de un brinco, fui a atisbar por un agujero entre las hojas y vi una humareda sobre el agua, algo más arriba..., casi delante del embarcadero. Y allí estaba el vapor, abarrotado de gente, flotando río abajo. Ya sabía qué pasaba. ¡Booom! Vi el humo blanco que salía del costado del vapor. ¿Saben? Estaban disparando cañonazos sobre el agua para sacar a flote mi cadáver.

Tenía hambre, pero no me convenía encender una hoguera, porque podrían ver el humo. Por lo tanto, permanecí sentado allí, contemplando el humo de los cañonazos. El río tenía en ese sitio una milla de ancho y siempre ofrecía un bonito aspecto en una mañana de verano, de modo que lo pasaba bastante bien viéndoles buscar mis restos, sólo que me acuciaba el hambre. Entonces se me ocurrió que siempre ponen azogue sobre hogazas de pan y las dejan flotar porque suelen localizar los cadáveres de los ahogados. Me dije que echaría un vistazo por si llegaba hasta mí alguna, y que daría yo buena cuenta de ella.

Me trasladé al extremo de la isla orientado hacia la orilla de Illinois para probar suerte y no quedé desfraudado. Se aproximó una enorme hogaza doble, y ya casi la había alcanzado con un palo largo, cuando me resbaló el pie y el pan siguió flotando alejándose de mí. Claro está que yo me encontraba donde la corriente era más intensa, lo sabía. Pero a poco llegó flotando otra hogaza de pan y esta vez me salí con la mía. Le saqué el tapón, sacudí la untadura de azogue y le hinqué los dientes. Era pan de tahona, del que come la gente bien, nada de vulgar pan de maíz.

Encontré un buen sitio entre las hojas y me senté allí, encima de un tronco, a comer el pan y a contemplar el vapor, sintiéndome muy satisfecho. Pensé algo. Me dije que seguramente la viuda, el cura o alguien estaría rogando que el pan me encontrase, y así había pasado. Por lo tanto, no había duda de que había algo de cierto en el asunto. Es decir, lo hay cuando una persona como la viuda o el cura rezan, pero eso no va conmigo, y supongo que sirve solamente para los buenos.

Encendí la pipa, eché una buena bocanada y continué mirando. El vapor flotaba con la corriente y calculé que podría tener la oportunidad de ver quiénes iban a bordo cuando se acercara, porque se aproximaría como la hogaza de pan. Cuando estuvo bastante cerca de mí, guardé la pipa, me encaminé al lugar donde había pescado el pan y me tumbé detrás de un tronco, en la ribera, en un sitio descubierto. Podía atisbar por la horquilla que formaba el tronco.

A poco se acercó y pasó tan cerca, que hubieran podido desembarcar bajando la pasarela de a bordo. Iba casi todo el mundo: papá, el juez Thatcher, Beckie Thatcher, Joe Harper, Tom Sawyer y su vieja tía

Polly, Sid, Mary y otras personas. Todos hablaban del crimen, pero el capitán los interrumpió diciendo:

—Fíjense bien: aquí es donde la corriente se acerca más, y tal vez le ha arrastrado hasta la orilla y está enredado en la maleza de la ribera. Así lo espero, de todos modos.

Yo no lo esperaba. Se agruparon todos y se apoyaron sobre las barandillas, casi mirándome a la cara, y estuvieron quietos, atentos, poniendo todo su empeño en escudriñar la orilla. Yo los veía muy bien, pero ellos a mí no. El capitán exclamó entonces:

—¡Apártense!

Y el cañón soltó una explosión tan horrible, que me ensordeció el ruido y casi me cegó el humo. Pensé que iba a morirme. Creo que si hubieran disparado con balas habrían conseguido el cadáver que andaban buscando.

Bueno, vi que no estaba herido, a Dios gracias. El vapor siguió flotando y desapareció de la vista detrás del cabo de la isla. Pude oír las detonaciones de vez en cuando, cada vez más distantes, y al cabo de una hora dejé de oírlas. La isla tenía tres millas de largo. Pensé que habían llegado a su final y que renunciaban a seguir buscándome. Pero me equivocaba. Doblaron el extremo de la isla y remontaron el canal por el lado del Missouri, con lentitud, disparando cañonazos ocasionalmente. Me dirigí a ese lado y los observé. Cuando estuvieron delante del cabo de la isla, dejaron de disparar, volvieron a la orilla del Missouri y regresaron al pueblo.

Supe entonces que estaba a salvo. Nadie vendría en mi busca. Saqué los bártulos de la canoa y me hice un campamento magnífico en la densa arboleda. Me levanté una especie de tienda con las mantas para resguardar las cosas de la lluvia. Capturé un pez y lo abrí con la sierra, y hacia la puesta del sol encendí la hoguera de campamento y cené. Después instalé una caña para pescar algo para el desayuno.

Cuando anocheció, me senté junto al fuego, fumando y sintiéndome bastante satisfecho, pero al rato me encontré algo solo, de modo que me dirigí a la orilla y escuché el rumor del agua, conté las estrellas, los troncos a la deriva y las balsas que descendían río abajo, y después me acosté: no hay mejor modo de pasar el tiempo cuando uno se siente solo; es imposible seguir estándolo, y entonces todo pasa.

Y así durante tres días y tres noches. No había diferencia..., siempre lo mismo. Pero al día siguiente fui a explorar la isla. Era su dueño, me pertenecía toda entera, por así decirlo, y quería saber cuanto a ella se refería, pero sobre todo quería pasar el tiempo. Encontré muchísimas fresas maduras, deliciosas, y uvas verdes de verano, y frambuesas, y ya empezaban a verse moras negras. «Pronto podría cogerlas para comerlas», pensé.

Fui de un lado para otro por los bosques hasta que calculé que estaba cerca de la isla. Llevaba la escopeta, pero no había disparado contra nada. La llevaba para protegerme, aunque mataría alguna pieza por la noche, cuando volviera a mi campamento. De momento estuve a punto de pisar una serpiente de buen tamaño, la cual se escurrió a través de la hierba y las flores, y yo fui tras ella intentando dispararle. Seguí avanzando y de repente me encontré justamente sobre las cenizas de una higuera de campamento aún humeante.

El corazón me dio un salto en el pecho. No esperé a mirar nada más, sino que desmartillé la escopeta y me largué por donde había venido, de puntillas, tan de prisa como pude. De vez en cuando me paraba un instante, entre la densa vegetación, y escuchaba, pero era tan fuerte mi respiración, que no podía oír nada. Seguí y algo más adelante volví a detenerme y escuché... Y así una y otra vez. Si veía un arbusto, lo tomaba por la figura de un hombre; si pisaba una ramita y la rompía, me daba la impresión de que me cortaban el aliento en dos y que mi mitad era la más pequeña.

No me sentía muy temerario cuando llegué al campamento, todo hay que decirlo, pero recordé que no era el momento de perder tiempo. Por lo tanto, recogí mis trastos para guardarlos en la canoa a fin de tenerlos fuera de la vista, apagué el fuego y desparramé las cenizas para darle el aspecto de una hoguera del año pasado. Después trepé a un árbol.

Creo que estuve allá arriba unas dos horas, pero no vi ni oí nada... Sólo me pareció ver y oír infinidad de cosas. Bueno, no podía quedarme allí arriba para siempre, de modo que acabé por bajar, aunque tuve buen cuidado de no salir de la espesura y de estar alerta todo el tiempo. No tenía para comer más que fresas y los restos del desayuno.

Cuando llegó la noche, me sentía realmente hambriento. Por eso, apenas me sentí protegido por las sombras, me alejé de la orilla antes de que saliera la luna y remé hacia la ribera de Illinois, a un cuarto de milla de distancia. Me adentré en los bosques y me preparé la cena. Había hecho el propósito de quedarme allí por la noche, cuando oí un plon-plon, plon-plon, y me dije: Se acercan caballos. En seguida oí voces. Lo puse todo dentro de la canoa lo más de prisa posible y me interné a rastras en el bosque para ver lo que podía encontrar. No había llegado lejos cuando oí decir a un hombre:

—Será mejor que acampemos aquí si encontramos un buen lugar; los caballos están rendidos. Echemos un vistazo por ahí.

No esperé y me apresuré a alejarme remando. Atraqué en el sitio de costumbre y decidí dormir dentro de la canoa.

No dormí mucho. Me lo impidió sobre todo el pensar. Y cada vez que me despertaba era con la sensación de que me cogían por el cuello. Por eso no me hizo mucho bien el sueñecito. Al fin llegué a decirme que así no podía vivir; tenía que descubrir quién estaba en la isla además de yo; lo averiguaría o reventaría. Y el caso es que en seguida me encontré mejor.

De manera que tomé mi pértiga para apartarme unos dos pasos de la orilla y luego dejé la canoa fuera arrastrada por la corriente hacia abajo, entre las sombras. Brillaba la luna, y más allá de las sombras estaba tan claro como si fuese de día. Fui escudriñándolo todo durante una hora: estaba silencioso y quieto como las rocas. Para entonces había llegado casi hasta un extremo de la isla. Se levantó una ligera brisa más bien fría y eso equivalía a decir que la noche casi había acabado. Giré con ayuda de la pértiga y apunté la canoa hacia el bosque. Me senté sobre un tronco y miré entre las hojas. Vi que la luna dejaba de vigilar y que las tinieblas empezaban a cubrir el río. Pero durante un ratito vi un haz pálido por encima de las copas de los árboles y comprendí que el día se acercaba.

De manera que cogí la escopeta y me deslicé hacía el lugar donde había encontrado el fuego de campamento, parándome cada dos minutos para escuchar. Pero no tuve suerte; no lo encontraba. Al fin vislumbré fuego a través de los árboles. Seguí su dirección despacio, cautelosamente. A poco me encontré lo bastante cerca para verlo, y allí había un hombre tendido en el suelo. Poco faltó para que me diera un ataque. Tenía una manta liada a la cabeza y ésta casi en el fuego. Me senté detrás de un grupo de arbustos, a unos seis pies de él, y miré con detenimiento. Ya lucía la claridad grisácea del nuevo día. El no tardó en bostezar, desperezarse y quitarse la manta... ¡Y era Jim, el negro de la señorita Watson! Apuesto a que me alegré de verle. Dije:

—¡Hola, Jim! —y salí de mi escondrijo.

El pegó un brinco y me miró con espanto. Cayó de rodillas y, juntando las manos, exclamó:

—¡No me hagas daño..., por piedad! Nunca he sido malo con un fantasma. Me agradan mucho los muertos y he hecho todo lo que he podido por ellos. Anda, entra de nuevo en el río, al sitio de donde vienes, y no le hagas nada al viejo Jim, que siempre será tu amigo...

Bueno, no tardé mucho en hacerle comprender que no estaba muerto. Me gustaba mucho ver a Jim. Ahora no estaba tan solo. Le dije que no tenía miedo de que él contara a la gente dónde estaba yo. Seguí hablando, pero él permanecía sentado, quieto, sin dejar de mirarme. Al fin le dije:

—Ya es de día. Vamos a desayunar. Reaviva tu fuego.

—¿De qué sirve reavivar el fuego para guisar frambuesas y hierbajos? Tienes escopeta, ¿no? Entonces podemos conseguir algo mejor que frambuesas.

—Entonces, ¿vives sólo de frambuesas y plantas? —le pregunté yo.

—No encontré otra cosa —contestó.

—Oye, ¿cuánto tiempo llevas en la isla, Jim?

—Llegué la noche después de que te mataran.

—¡Caramba! ¿Tanto tiempo?

—Sí, sí...

—¿Y no has tenido más que esa porquería para comer?

—No..., nada más.

—Bueno, pues debes estar muerto de hambre, ¿verdad?

—Creo que sería capaz de comerme un caballo. ¿Desde cuándo estás en la isla?

—Desde la noche de mi asesinato.

—¡No! Oye, ¿de qué has vivido? Pero ¿tienes escopeta? ¡Oh, claro, tienes escopeta! Eso va bien. Ahora mata algo y yo me ocupo del fuego.

Entonces fuimos a donde estaba la canoa y, mientras él encendía una hoguera en un claro cubierto de hierba, yo traje comida, tocino, café, una cafetera y una sartén, azúcar y tazas de hojalata, y el negro se impresionó de lo lindo porque supuso que lo conseguía por artes mágicas. También capturé un buen pez y Jim lo limpió con su cuchillo y lo frió.

Cuando el desayuno estuvo listo, nos tumbamos en la hierba y comimos comida caliente. Jim la atacó con todas sus fuerzas, porque estaba casi muerto de hambre. Luego, cuando nos encontramos bastante hartos, estuvimos echados desperezándonos.

Al poco rato dijo Jim:

—Pero oye, Huck, ¿a quién mataron en la cabaña si no eras tú?

Se lo conté todo, y Jim dijo que había sido una idea inteligente. Dijo que ni Tom Sawyer habría pensado en un plan mejor. Después le pregunté:

—¿Cómo y por qué has venido aquí, Jim?

Pareció bastante confuso y durante un minuto no dijo nada. Al fin contestó:

—Tal vez sea mejor no decirlo.

—¿Por qué, Jim?

—Verás, hay razones. Pero tú no me delatarás, si te lo cuento, ¿verdad, Huck?

—¡Que me cuelguen si lo hago, Jim!

—Bueno, creo en ti, Huck... ¡Me escapé!

—¡Jim!

—Oye, dijiste que no me delatarías... Recuerda que lo has dicho, Huck.

—Pues sí, lo dije y cumpliré mi palabra. Palabra de indio injun que no lo diré. La gente me llamaría ruin abolicionista y me despreciaría por callarme..., pero eso no cambia las cosas. No lo diré ni volveré allá de ninguna manera. Así que cuéntamelo todo.

—Pues, verás, fue así: La vieja solterona..., es decir la señorita Watson, me hace la vida imposible, me trata bastante mal, pero siempre dijo que me vendería en Orleans. Me di cuenta de que últimamente rondaba por allí a menudo un traficante de negros y empecé a preocuparme. Bueno, pues una noche me llegué hasta la puerta, que no estaba bien cerrada, oí cómo la solterona decía a la viuda que pensaba venderme en Orleans, aunque no quería, pero que podría conseguir ochocientos dólares por mí y que no podía resistirse ante tanto dinero. La viuda trató de convencerla para que no me vendiera, pero yo no esperé a enterarme del resto. Te aseguro que salí huyendo más que de prisa.

No paré de correr colina abajo. Esperaba robar un esquife más arriba del pueblo, pero aún había gente levantada y me escondí en la tonelería de la orilla para esperar a que se fueran. Pasé allí la noche entera. Siempre había alguien rondando por aquel lugar. A las seis de la mañana empezaron a pasar esquifes, y a las ocho o las nueve todos los que iban en los esquifes decían que tu papá había llegado al pueblo diciendo que te habían matado. Los últimos esquifes iban llenos de damas y caballeros, que se dirigían al sitio para verlo. A veces se detenían en la orilla a descansar antes de atravesar el río y así, por lo que hablaron, me enteré del crimen. Sentí mucho que te mataran, Huck, pero ahora estoy contento.

Pasé todo el día tumbado debajo de las bacías. Tenía hambre, pero no estaba asustado porque sabía que la solterona y la viuda irían a pasar todo el día a la reunión en un campamento después de desayunar, y ellas saben que yo salgo con el ganado al amanecer, de modo que no esperarían verme allí y no me echarían de menos hasta la noche. Los otros criados no me echarían de menos, porque se largarían en cuanto se marcharan las viejas.

Pues bien, cuando fue de noche, tomé el camino del río y recorrí unas dos millas o más hasta que no había casas. Ya había decidido lo que iba a hacer. Date cuenta de que, si continuaba huyendo a pie, los perros descubrirían mi pista. Si robaba un esquife para cruzar el río, lo echarían de menos y sabrían que habría desembarcado en la otra orilla y descubriría mi pista. Entonces me dije que lo mejor sería encontrar una balsa; una balsa no deja rastro.

Vi una luz que doblaba el cabo, y entonces nadé apoyándome en un tronco hasta más de la mitad del río, me mezclé entre los demás troncos a la deriva, bajando la cabeza, y nadé contra corriente hasta que llegó la balsa. Entonces nadé hasta la popa y me agarré a ella. Se nubló y todo estuvo bastante oscuro un rato, de modo que me encaramé a las tablas y me tumbé encima. Los hombres estaban en el centro, junto a una linterna. El río estaba crecido y la corriente era bastante fuerte, de modo que pensé que alrededor de las cuatro de la mañana estaría veinticinco millas más abajo del río y que entonces me escabulliría, antes de clarear el día, llegaría a nado hasta la orilla y me adentraría en los bosques por el lado de Illinois.

Pero no tuve suerte. Cuando estábamos casi en el cabo de la isla, un hombre se acercó con la linterna. Vi que era inútil esperar más, así que me escurrí al agua por la borda y nadé hacia la isla. Bueno, tenía la idea de que podría tomar tierra en cualquier parte, pero fue imposible allí, porque la ribera era demasiado escarpada. Anduve luego por casi toda la isla antes de encontrar un buen sitio. Entré en el bosque y pensé que ya no haría más el tonto con las balsas mientras movieran tanto la linterna. Tenía la pipa, un poco de tabaco y algunas cerillas en mi gorra, y no estaban mojadas; por lo tanto, me di por satisfecho.

—¿Y no has tenido pan ni carne para comer durante todo ese tiempo? ¿Por qué no cogías tortugas?

—¿Y cómo iba a atraparlas? ¡No es tan fácil agarrarlas! ¿Y cómo puede uno golpearlas con una piedra? ¿Cómo hacerlo por la noche? Porque, como puedes suponer, no iba a aparecer durante el día en la orilla del río.

—Sí, claro... Has tenido que ocultarte en el bosque todo el tiempo, claro. ¿Les oíste disparar los cañonazos?

—¡Oh, sí! Sabía que te buscaban a ti. Les vi pasar por aquí cerca y los estuve mirando desde los arbustos.

Se acercaron algunos pájaros pequeños, volando una o dos yardas y volviendo a posarse. Jim dijo que era señal de que iba a llover. Así pasaba cuando las gallinas volaban de esta forma, por lo que sucedería lo mismo con los pájaros. Jim no me dejó que cogiera algunos, y afirmó que hacerlo atraía la muerte. Dijo que una vez, estando enfermo su padre, algunas personas capturaron un pájaro, y entonces su vieja abuela dijo que su padre moriría, y murió.

Y Jim aseguró que no debían contarse las cosas que uno iba a preparar para comer, porque traía mala suerte. Lo mismo ocurría si uno sacudía el mantel después de ponerse el sol. Y dijo que, si un hombre era dueño de una colmena y se moría, había que decírselo a las abejas antes de que saliera el sol al día siguiente, porque de lo contrario las abejas se debilitaban, dejaban de trabajar y se morían. También dijo Jim que las abejas no picaban a los idiotas, pero esto no lo creí, porque yo las he desafiado a picarme muchas veces y nunca lo han hecho.

Había oído algunas cosas parecidas antes, pero no todas. Jim conocía toda clase de señales. Dijo que lo sabía casi todo. Yo afirmé que a mí me parecía que todas las señales indicaban mala suerte y le pregunté que si no había señales de buena suerte. El contestó:

—Muy pocas... y no son útiles a nadie. ¿Para qué quieres saber cuándo te llegará la buena suerte? ¿Quieres ahuyentarla? —Y añadió—: Si tienes los brazos y el pecho peludos, es señal de que serás rico. Bueno, pues una señal como ésta sí es útil, porque es adelantada. Mira, puede que seas pobre mucho tiempo y entonces podrías desanimarte y matarte, si no creyeras que esa señal quiere decir que serás rico.

—¿Y tú tienes los brazos y el pecho peludos, Jim?

—¿Por qué haces esa pregunta? ¿No ves que sí?

—Bueno, ¿y eres rico?

—No, pero lo fui y volveré a ser rico. Una vez tuve catorce dólares, pero me metí a especular y me arruiné.

—¿En qué especulabas, Jim?

—Bueno, primero probé con mercancías.

—¿De qué clase?

—Pues mercancías vivas. Ganado, ¿entiendes? Invertí diez dólares en una vaca. Pero no volveré a exponer dinero en mercancías. La vaca se murió en mis manos.

—De modo que perdiste los diez dólares.

—No, no los perdí. Unicamente perdí nueve. Vendí el pellejo por un dólar y diez centavos.

—Te quedaron cinco dólares y diez centavos. ¿Volviste a especular?

—Sí, ¿conoces al negro cojo que pertenece al viejo señor Bradish? Bueno, pues abrió un banco, y a todo el que ponía un dólar le daba cuatro dólares más al cabo de un año. Bien, todos los negros lo hicieron, pero tenían poco. Yo era el que tenía más. De modo que pedí más de cuatro dólares, y le dije que si no me los daba abriría yo mismo otro banco. Claro, ese negro quería quitarme del negocio, porque dijo que no había bastante para dos bancos, de modo que le dije que pondría mis cinco dólares y él me pagaría treinta y cinco al cabo del año.

Así lo hice. Entonces pensé que invertiría los treinta y cinco dólares para seguir especulando. Había un negro llamado Bob, que había cogido una barca chata y su amo no lo sabía. Se la compré y le dije que le pagaría los treinta y cinco dólares al finalizar el año, pero esa noche robaron la barca y al día siguiente el negro cojo dijo que el banco había quebrado, de modo que ninguno de nosotros sacó dinero.

—¿Qué hiciste con los diez centavos, Jim?

—Iba a gastarlos, pero tuve un sueño, y en el sueño me decían que los diera a un negro llamado Balum... El «asno de Balum» le llamaban, para acortar la historia. Es uno de ésos que están mal de la cabeza, ¿sabes? Pero dicen que es afortunado, y yo vi que yo no lo era. El sueño decía que dejara que Balum invirtiera los diez centavos y él los aumentaría por mí. Bueno, Balum cogió el dinero y, cuando estaba en la iglesia, oyó decir al predicador que aquél que daba a los pobres hacía un préstamo a Dios, y que recibiría su dinero cien veces doblado. De modo que Balum dio los diez centavos para los pobres y se tumbó a esperar lo que saldría de aquello.

—Bueno, ¿y qué pasó, Jim?

—Nada, nada en absoluto. No hubo modo de cobrar aquel dinero, ni tampoco pudo Balum. No prestaré más dinero como no sea sobre seguro. ¡Mira que decir el predicador que el dinero sería devuelto cien veces doblado! Si yo lograra que me devolvieran los diez centavos, lo encontraría justo y me alegraría de la operación.

—Bien, de todos modos, no pasa nada, porque un día u otro serás rico, Jim.

—Sí..., lo soy ahora, mirándolo bien. Soy mi dueño y valgo ochocientos dólares. Ojalá tuviera ese dinero, ya no querría más.

9

Yo quería ir a ver un lugar en el centro de la isla, que encontré mientras exploraba; nos pusimos en marcha y no tardamos en llegar, porque la isla sólo tenía tres millas de largo y un cuarto de milla de ancho.

Ese sitio era una colina o sierra de unos cuarenta pies de altura. Nos costó bastante subir arriba, tan abruptas eran las laderas y tan espesa la maleza. La exploramos bien por todas partes y al poco rato encontramos una enorme caverna en la roca, casi en la cima, en el lado orientado hacia Illinois. La caverna ocupaba como dos o tres habitaciones unidas y Jim podía estar de pie en su interior. Dentro hacía fresco. Jim propuso guardar allí nuestras cosas, pero yo dije que no, porque tendríamos que subir y bajar a cada momento.

Jim aseguró que, si tuviéramos la canoa escondida en un buen sitio y todas las cosas en la caverna, podríamos refugiarnos allí en el caso de que alguien viniera a la isla, y que nunca nos encontrarían sin los perros. Además, dijo que los pajaritos habían señalado que iba a llover, y ¿quería yo que se empapara todo?

De modo que volvimos, la llevamos a remo hasta delante de la caverna y subimos allí todos nuestros bártulos. Luego buscamos un sitio cercano donde ocultar la canoa entre los sauces. Sacamos algunos peces de los anzuelos, que volvimos a dejar preparados. Entonces guisamos la comida.

La entrada de la caverna era lo bastante ancha para hacer rodar por ella un barril, y a un lado de la entrada el suelo sobresalía un poco, era llano y resultaba ideal para encender una hoguera, de modo que allí preparamos la cena.

Dentro extendimos las mantas para alfombrarlo y comimos allí. Dejamos las otras cosas en el fondo de la caverna. Pronto oscureció y empezó a tronar y relampaguear. Los pájaros habían acertado. En seguida empezó a llover con toda la furia y nunca había visto un viento tan fuerte. Era una tormenta normal de verano. Estaba tan oscuro, que fuera parecía todo de color azul-negro y resultaba encantador; la lluvia azotaba con tal fuerza los árboles cercanos, que parecían telarañas. De pronto llegaba una ráfaga de viento que doblaba los árboles y daba la vuelta a las hojas, mostrando su palidez de la parte inferior. Seguía después un verdadero ramalazo huracanado, que obligaba a los árboles a extender sus ramas como si se volvieran locos. Después, cuando reinaba la oscuridad más azulada..., clareaba con el brillo de la gloria y se vislumbraban las copas de los árboles revueltas, más allá, en medio de la tormenta, a centenares de yardas más lejos de lo que llegaba la vista, negro como el pecado en un segundo; y ahora se podía oír el trueno con todo su fragor y alejarse arrancando ecos de furia y gruñendo, rodando por el cielo hacia el interior del mundo como ruedan los barriles vacíos por una escalera, cuando ésta es larga y rebotan muchísimo, ¿comprenden?

—Es bonito esto, Jim —dije yo—. No querría estar en otro sitio. Pásame otro trozo de pescado y pan de maíz caliente.

—Bien, pues no estarías aquí si no es por Jim. Ahora estarías allá abajo, en el bosque, sin comida, casi ahogándote bajo este diluvio. No te quepa la menor duda, chico. Las gallinas saben cuándo va a llover y también los pájaros, amigo.

El río siguió creciendo y creciendo durante diez o doce días, hasta que al fin desbordó las orillas. En los lugares bajos de la isla el agua tenía tres o cuatro pies de profundidad, así como en la parte baja del Illinois. En aquella parte tenía una anchura de muchas millas, pero en la del Missouri había la misma, de una media milla, porque la orilla del Missouri no era más que una muralla de altos acantilados.

Algunas veces atravesábamos la isla en la canoa. En los profundos bosques había frescor y sombra, aunque fuera llamease el sol. Serpenteábamos saliendo y entrando por los bosques, y en ocasiones las enredaderas colgantes eran tan espesas que teníamos que volver atrás y seguir por otro camino. Bueno, pues en todos los árboles viejos y destruidos podían verse conejos, serpientes y otros animales; y cuando la isla estuvo inundada, durante un par de días, el hambre los hizo tan dóciles, que podíamos acercarnos a remo y ponerles la mano encima, pero no a las serpientes y las tortugas, que se escurrían dentro del agua. Si hubiéramos querido, habríamos tenido muchísimos animalitos domésticos.

Una noche recogimos una pequeña parte de una balsa de troncos. Tenía doce pies de ancho por unos quince o dieciséis de largo, y la parte superior se alzaba sobre el agua unas seis o siete pulgadas, formando un piso sólido y plano. A veces veíamos pasar troncos del aserradero durante el día, pero los dejábamos; no queríamos hacernos ver a la luz del día.

Otra noche, cuando nos encontrábamos en la parte superior de la isla, poco antes del amanecer, vimos descender una casa de madera por el lado oeste. Tenía dos pisos y se balanceaba mucho. Nos acercamos a ella remando y subimos a bordo, trepando hasta una ventana del piso de arriba. Pero era demasiado oscuro para verse, de modo que amarramos la canoa y nos sentamos dentro a esperar que fuera de día.

Empezó a amanecer antes de que llegáramos a la isla. Entonces nos asomamos por la ventana. Vimos una cama, una mesa y dos sillas viejas, y montones de cosas por doquier, en el suelo. Y había ropas colgadas en la pared. En el suelo, en el extremo opuesto yacía algo que parecía un hombre. Jim dijo:

—¡Eh, oiga!

Pero no se movió. Entonces yo grité también, y Jim dijo:

—Ese hombre no duerme... Está muerto. Quédate aquí quieto, yo entraré a ver.

Penetró en el cuarto, se inclinó a su lado y exclamó:

—¡Está muerto! Y además desnudo. Le dispararon por la espalda. Me parece que lleva dos o tres días muerto. Entra, Huck, pero no le mires la cara... Es espantosa.

No le miré. Jim le cubrió con algunos trapos viejos, pero no había necesidad de que lo hiciera. No quería verlo. Por el suelo había muchas cartas grasientas desparramadas, botellas de whisky vacías y un par de antifaces de tela negra. Y las paredes estaban cubiertas de palabras y dibujos hechos con carbón. Había dos vestidos viejos y sucios de calicó, un sombrero de sol y ropa interior femenina colgados en la pared, así como ropas de hombre. Lo metimos todo en la canoa. Podía ser útil. Había un viejo sombrero de paja jaspeada de muchacho en el suelo. También me lo llevé. Y una botella que había contenido leche y que tenía un tapón de trapo para que lo chupara un niño. Nos habríamos llevado la botella si no hubiera estado rota. Había un arca vieja y destrozada y un baúl de pelo con los goznes rotos. Estaban abiertos, pero dentro no había nada interesante. Había tal desorden, que pensamos que la gente se había marchado precipitadamente sin poder llevarse todos sus efectos.

Cogimos una linterna vieja de hojalata y un cuchillo de carnicero sin mango, un cuchillo «Barlow» nuevo, que valdría cincuenta centavos en cualquier almacén y un montón de velas de sebo, un candelero de hojalata, una calabaza vinatera y una taza de hojalata; además, una vieja colcha de la cama y una bolsa con alfileres, agujas de coser, cera y botones, hilo y cosas por el estilo, así como una hacheta y algunos clavos, una caña de pescar tan gruesa como mi dedo, con algunos anzuelos monstruosos, un rollo de ante y un collar de cuero, de perro, una herradura de caballo y algunas redomas de medicinas sin etiquetas. Y cuando ya nos íbamos encontré una almohaza en bastante buen estado y Jim halló un viejo arco de violín y una pata de palo. Tenía las correas rotas, pero, exceptuando este detalle, era una buena pata de palo, aunque resultaba demasiado larga para mí y demasiado corta para Jim. No pudimos encontrar la otra aunque lo registramos todo.

Así que, en conjunto, conseguimos un buen botín. Cuando estábamos listos para irnos, nos hallábamos a un cuarto de milla más abajo de la isla y era completamente de día, por lo que hice que Jim se echara en el fondo de la canoa y se tapara con la colcha, ya que si permanecía sentado la gente habría visto de lejos que era un negro. Remé hacia la orilla del Illinois y fuimos a la deriva más de media milla. Me acerqué al agua mansa al pie de la ribera y no tuve accidentes ni vi a nadie. Llegamos a casa a salvo.

10

Después de desayunar quise hablar del muerto y hacer cábalas sobre cómo se había producido su asesinato, pero Jim se negó a comentarlo. Dijo que traería mala suerte. Además, que su espíritu podría venir a torturarnos. Afirmó que era más probable que un hombre que no había sido enterrado volviera a molestar, que otro confortablemente sepultado. Eso parecía razonable, de modo que no insistí más, pero no pude por menos de recordarlo, deseando saber quién había matado de un disparo al hombre y por qué lo hizo.

Revolvimos las ropas que nos habíamos llevado y encontramos ocho dólares de plata cosidos en el forro de un viejo abrigo de lana. Jim supuso que la gente de aquella casa lo había robado, porque, si llegan a saber que el dinero estaba allí, no lo habrían dejado. Dije que a mí me parecía que le habían matado, pero Jim no quiso hablar de ello, y yo agregué:

—Bueno, tú crees que trae mala suerte, pero ¿qué dijiste cuando te traje la piel de serpiente que encontré en la cima de la sierra anteayer?

Dijiste que tocar una piel de serpiente con las manos es la peor mala suerte del mundo. Bueno, ¡aquí tienes la mala suerte! Hemos conseguido todas esas cosas y además ocho dólares. ¡Ojalá tuviéramos tan mala suerte como todos los días, Jim!

—No importa, amigo, no importa. No te las prometas tan felices. Vendrá. Te digo que vendrá.

Y vino. Tuvimos esa charla el martes. Bueno, pues el viernes, después de comer, estábamos tumbados en la hierba en lo alto de la loma y nos quedamos sin tabaco. Bajé por él a la caverna y allí encontré una serpiente. La maté y la dejé enroscada a los pies de la manta de Jim, como si estuviera viva, pensando que resultaría divertido ver lo que pasaba cuando Jim la viera. Por la noche me había olvidado por completo de la serpiente y, cuando Jim se echó sobre la manta mientras yo encendía una vela, encontró la pareja de la serpiente muerta, que le mordió.

Jim se levantó dando un alarido, y lo primero que iluminó la luz fue el reptil enroscado y erguido, dispuesto a lanzarse al segundo ataque. Lo aplastó en un segundo y Jim cogió el barril de whisky de papá y empezó a vaciarlo.

Iba descalzo y la serpiente le había mordido en el talón. Eso pasó por ser yo un estúpido al no recordar que allí donde uno deja una serpiente muerta viene su pareja y se enrosca a su alrededor. Jim me dijo que cortara la cabeza de la serpiente y la arrojara lejos, y después le desollara el tronco y asara un trozo. Lo hice y Jim se lo comió diciendo que eso le ayudaría a curarse. Me hizo arrancarle los cascabeles y atarlos alrededor de su muñeca. Aseguró que eso ayudaría. Luego salí y lancé lejos a las serpientes, entre los arbustos, porque no quería que Jim se enterase, si yo podía evitarlo, de que la culpa había sido mía.

Jim bebió whisky y más whisky del barril, y de vez en cuando le daba un ataque de locura y se tiraba al suelo aullando, pero en cuanto volvía en sí se agarraba otra vez al barril de whisky. Tenía el pie muy hinchado y también la pierna, pero poco a poco la borrachera hizo su aparición y pensé que Jim ya estaba bien, pero yo hubiera preferido ser mordido por una serpiente, que por el whisky de papá.

Jim estuvo acostado durante cuatro días con sus noches. Luego la hinchazón desapareció y él volvió a andar. Hice el propósito de no volver a coger nunca más la piel de una serpiente con las manos, después de comprobar los resultados. Jim dijo que esperaba que la próxima vez le hiciera caso. Añadió que manosear una piel de serpiente acarreaba una mala suerte tan espantosa, que quizás no pararía ahí la racha. Dijo que prefería mil veces ver la luna nueva por encima de su hombro izquierdo, antes que coger una piel de serpiente con la mano. Y yo empecé a pensar lo mismo, aunque siempre he creído que mirar la luna nueva por encima del hombro izquierdo es una de las cosas más temerarias e insensatas que uno puede hacer. El viejo Hank Bunjer lo hizo una vez y se jactó de ello, pero antes de los dos años se emborrachó, y se cayó desde una torre y quedó de tal manera, que parecía, como quien dice, un amasijo de carne desparramada; y lo metieron de lado entre dos puertas de granero a guisa de ataúd y lo enterraron de esta manera, según cuentan, pero yo no lo vi. Papá me lo contó. Pero, en todo caso, eso le pasó por loco, por mirar a la luna de aquel modo.

Pasaron los días y el río volvió a su cauce entre las márgenes; una de las primeras cosas que hicimos fue poner de cebo un conejo desollado en uno de los enormes anzuelos y capturar un pez tan grande como un hombre, de unos seis pies y dos pulgadas de largo y de más de doscientas libras de peso. No pudimos atraparlo con las manos, naturalmente, pues nos hubiera arrojado de cabeza al Illinois. Nos quedamos sentados allí, viéndolo agitarse y moverse hasta que se ahogó. Encontramos un botón de latón dentro de su estómago y una pelota redonda, así como muchas porquerías. Partimos la pelota por la mitad y dentro había una canilla. Jim dijo que debía haberlo llevado dentro mucho tiempo para tenerla cubierta de tal manera, formando una pelota. Creo que fue el mayor pez que se capturó en el Mississippi. Jim aseguró no haber visto nunca nada igual. En el pueblo valdría una buena suma. En su mercado venden pescado como éste a tanto la libra; todo el mundo lo compra; su carne es tan blanca como la nieve y hace una deliciosa fritura.

Al día siguiente dije que empezaba a encontrar aquella vida monótona y aburrida, y que necesitaba alguna emoción nueva. Dije que pensaba cruzar el río para averiguar qué pasaba en la orilla opuesta. A Jim le sedujo la idea, pero observó que había que ir de noche y estar muy alerta. Luego lo estudió y preguntó por qué no me ponía algunas de las ropas viejas y me disfrazaba de chica. También era una excelente idea. Así que acortamos uno de los vestidos de calicó, me arremangué los pantalones hasta las rodillas y me lo puse. Jim me lo abrochó por la espalda con los corchetes y no me sentaba del todo mal. Me ajusté el sombrero de sol atándolo por debajo de la barbilla, así que para quien quisiera verme la cara era como mirar por la tubería de una estufa. Jim dijo que nadie me conocería, salvo de día, y aun entonces a duras penas. Estuve todo el día ensayando para andar con aquellas cosas puestas y lo conseguí al fin, aunque Jim dijo que no sabía andar como una chica, añadiendo que debía dejar la costumbre de levantarme las faldas para meter la mano en el bolsillo de mis pantalones. Tomé buena nota del detalle y lo hice mejor.

Poco después de anochecer me dirigí hacia la orilla de Illinois. Crucé hacia el pueblo desde un poco más abajo del embarcadero del vapor y, llevado por la corriente, llegué a la orilla del pueblo. Amarré y empecé a andar a lo largo de la ribera. Había luz en una cabaña pequeña que estuvo deshabitada durante mucho tiempo, y me pregunté quién viviría entonces allí. Me aproximé sigilosamente y atisbé por la ventana. Dentro había una mujer de unos cuarenta años de edad, haciendo calceta al lado de una vela, que estaba encima de una mesa de pino. Su rostro me era desconocido. Debía ser forastera, porque no había ninguna cara en el pueblo que no me fuera familiar. Fue un golpe de suerte, porque ya empezaba a acobardarme. Tenía miedo de haber venido. La gente podía reconocer mi voz y descubrirme. Pero si aquella mujer había estado dos días en un pueblo tan pequeño, podría contarme lo que yo quería saber; por lo tanto, llamé a la puerta con el firme propósito de recordar que era una chica.

11

—¡Adelante! —exclamó la mujer, y entré; luego dijo—: Toma una silla.

Obedecí. Me miró de cuerpo entero con sus ojillos relucientes y añadió:

—¿Cuál es tu nombre?

—Me llamo Sarah Williams.

—¿Dónde vives? ¿En este vecindario?

—No, señora, en Hookerville, siete millas más abajo. He venido andando y estoy agotada.

—Y hambrienta, supongo. Te prepararé algo.

—No, no tengo hambre. Tenía tanta, que tuve que detenerme dos millas más abajo en una granja, y por eso ya no tengo hambre. Por esto llego tan tarde. Mi madre está enferma, sin dinero y sin nada, y vengo a decírselo a mi tío Abner Moore. Vive en la parte alta del pueblo, dice mamá. Nunca he estado aquí. ¿Le conoce usted?

—No, pero es que aún no conozco a todo el mundo. Vivo aquí desde hace dos semanas. Hay mucho trecho desde aquí a la parte alta del pueblo. Será mejor que pases aquí la noche. Quítate el sombrero.

—No —dije yo—, descansaré un ratito y me iré. No me da miedo la oscuridad.

Dijo que no me dejaría marcharme sola, que su marido llegaría dentro de una hora y media y él me acompañaría. Luego empezó a hablar de su marido, de sus familiares, que vivían río arriba y de otros familiares que vivían río abajo, de la buena posición en que estaban antes y de que les parecía que habían cometido una equivocación viniendo a nuestro pueblo en lugar de dejar lo seguro..., y así una y otra vez, hasta que llegué a sospechar que era yo quien se había equivocado entrando en la casa para saber qué pasaba en el pueblo. Poco a poco empezó a hablar de papá y del crimen, y me pareció de perlas que charlara por los codos. Me habló de mí y de Tom Sawyer, cuando encontramos los seis mil dólares (sólo que ella dijo diez), de papá, de su mala reputación y de la mía, también mala, y finalmente llegó al asunto de mi asesinato. Entonces yo dije:

—¿Quién lo hizo? Hemos oído muchas cosas sobre esto allá abajo, en Hookerville, pero no sabemos quién mató a Huck Finn.

—Bueno, supongo que aquí hay más de uno que quisiera saber quién lo mató. Algunos creen que lo hizo el propio Finn, su padre.

Papá estaba de pie a mi lado, con aire torvo y aspecto enfermizo. Dijo:
—¿Qué haces con esa escopeta? (pág. 30).

—No... ¿Es posible?

—Así lo creyó casi todo el mundo al principio. El nunca sabrá cómo se libró de ser linchado. Pero antes de anochecer opinaron lo contrario y decidieron que lo hizo un negro fugitivo que se llamaba Jim.

—¡El...!

Me callé. Comprendí que me convenía callar. Ella continuó, sin darse cuenta siquiera de mi interrupción.

—El negro se escapó la misma noche en que mataron a Huck Finn. Ahora ofrecen una recompensa por él... Trescientos dólares. Y también hay una recompensa por el viejo Finn... Doscientos dólares. Verás, él vino al pueblo a la mañana siguiente del crimen para contarlo, y fue con los demás en el vapor y después desapareció. Antes de que anocheciera querían lincharlo, pero se había ido, ¿comprendes? Bueno, al día siguiente descubrieron que el negro se había escapado. Se supo que no le había visto nadie desde las diez de la noche del crimen. Le cargaron a él el crimen, ¿sabes? Y entonces, al día siguiente, regresó el viejo Finn y acudió gimoteando al juez Thatcher pidiéndole dinero para capturar al negro por todo Illinois. El juez le dio dinero y por la noche el viejo Finn se emborrachó y estuvo rondando por ahí hasta después de medianoche con un par de forasteros de aspecto sospechoso, con quienes después se marchó. Bueno, pues todavía no ha vuelto, y nadie espera verlo de nuevo hasta que se olvide ese asunto, ya que la gente cree que él fue quien mató a su hijo y arregló las cosas para que todos creyeran que lo habían hecho los ladrones y conseguir así el dinero de Huck sin tener que molestarse con un largo juicio. Dice la gente que era lo bastante canalla para hacerlo. ¡Oh!, supongo que es astuto. Si no regresa en un año, estará salvado. No se le puede probar nada, ¿sabes? Entonces los ánimos estarán menos exaltados y se apoderará del dinero de Huck con la mayor facilidad.

—Sí, eso supongo, señora. No veo ningún obstáculo para que no lo haga. ¿Ha dejado de pensar todo el mundo que lo hizo el negro?

—¡Oh, no; no todos! Hizo muchas cosas. Pero pronto atraparán al negro, y puede que le hagan hablar metiéndole el miedo en el cuerpo.

—Pero, ¿todavía lo persiguen?

—¡Vaya! ¡Si serás cándida, muchacha! ¿Te parece que se encuentran todos los días trescientos dólares? Algunos creen que el negro no anda lejos de aquí. Yo soy de éstos..., pero no lo digo. Hace algunos días estaba hablando con un anciano matrimonio que vive en la casa de al lado, en la cabaña de troncos, y dijeron que nadie va a esa isla que hay más allá, la isla Jackson. ¿No vive nadie allí?, pregunté yo. «Nadie», me contestaron. No dije más, pero pensé mucho. Estaba casi segura de haber visto humo allá, en la parte del cabo, un par de días antes, de modo que me dije a mí misma que seguramente el negro se esconde allí. Al menos vale la pena registrar el lugar. Desde entonces no he visto humo, y me figuro que se ha ido, si es que era él, pero mi marido irá a comprobarlo... con otro hombre. Estuvo ausente, río arriba, pero ha regresado hoy, y se lo dije en cuanto llegó, hace dos horas.

Me puse tan nervioso que no podía estarme quieto. Tenía necesidad de ocupar las manos en algo, de manera que cogí una aguja de encima de la mesa y empecé a enhebrarla. Me temblaban las manos y no me salía muy bien. Cuando la mujer dejó de hablar, levanté los ojos y vi que me miraba con mucha curiosidad, sonriendo ligeramente. Dejé la aguja y el hilo y demostré interés —y lo tenía desde luego— diciendo:

—Trescientos dólares es mucho dinero. ¡Ojalá los tuviera mi madre! ¿Irá esta noche su marido?

—¡Sí! Subió al pueblo con el hombre de quien te he hablado para conseguir un bote y ver si les prestan otra arma. Irán después de medianoche.

—¿No lo verían mejor si esperasen a que fuera de día?

—Sí, pero ¿no podría ver mejor también el negro? Seguramente después de medianoche estará durmiendo y ellos podrán explorar los bosques en busca de la hoguera de su campamento, si lo tiene.

—No se me había ocurrido.

La mujer seguía mirándome con curiosidad, por lo que me sentí muy violento. A poco dijo ella:

—¿Cómo dijiste que te llamas, preciosa?

—Ma... Mary Williams.

Tuve la sensación de que no había dicho Mary Williams antes, de modo que mantuve bajos los ojos. Me parecía haber dicho Sarah Williams. Me sentí como acorralado y tenía miedo de demostrarlo. Deseé que la mujer dijera algo más. Cuanto más rato estaba callada, tanto más intranquilo me sentía yo. Pero al fin dijo:

—Cariño, me parece que al entrar dijiste que te llamabas Sarah.

—¡Oh, sí, lo dije, señora! Sarah Mary Williams. Algunos me llaman Sarah y otros Mary.

—¡Ah!, ¿de veras?

—Sí, señora.

Me encontraba mejor, pero deseé estar fuera de todos modos. Todavía me faltaba valor para levantar los ojos.

Bueno, la mujer empezó a hablar de los malos tiempos que corrían, de lo pobremente que vivían y de que las ratas iban de un lado a otro con tanta libertad como si la casa fuera suya, y así sucesivamente, y yo recobré al fin la calma. Tenía razón en lo de las ratas. De vez en cuando se veía una asomando el hocico por un agujero en el rincón. Ella dijo que tenía que tener cosas a mano para tirarlas a las ratas cuando se quedaba sola, porque de lo contrario no estaría tranquila. Me mostró una barra de plomo retorcida y formando un nudo, dijo que solía tener buena puntería con ella, pero que unos días atrás se había dislocado el brazo y no sabía si ahora podría acertar. Aguardó una oportunidad y ágilmente lanzó la barra contra una rata, errando con mucho el tiro. Profirió un «¡huy!» de dolor. Entonces me pidió que yo probara la próxima vez.

Yo quería marcharme antes de que regresara su marido, pero no lo demostré, claro está. Cogí la barra y a la primera rata que asomó el hocico se la tiré. La hubiera dejado bastante mal herida si no se hubiera movido de donde estaba. Dijo la mujer que había sido un lanzamiento de primera y me brindó una segunda oportunidad. Fue a recoger la barra de plomo y trajo también una madeja de hilo con la que quería que yo la ayudase. Sostuve en alto las dos manos y ella puso encima la madeja y continuó hablando de sí misma y de los asuntos de su marido. Pero se interrumpió para decir:

—Vigila las ratas. Será mejor que tengas la barra de plomo a mano, en el regazo.

Así que soltó la barra en mi regazo, yo junté las piernas para sostenerla y ella siguió charlando. Pero sólo durante un minuto. Luego quitó la madeja, me miró fijamente a los ojos con expresión afable y preguntó:

—Veamos, ahora... ¿Cómo te llamas de verdad?

—¿Co... cómo, señora?

—¿Cuál es tu nombre? ¿Bill, Tom, Bob... o qué?

Creo que yo temblaba como una hoja. No sabía qué hacer. Pero dije:

—Por favor, no se burle de una pobre chica como yo, señora. Si le estorbo en su casa, yo...

—No lo harás. Siéntate y quédate donde estás. No voy a hacerte daño ni tampoco te delataré. Debes contarme tu secreto y confiar en mí. Lo guardaré y, además, te ayudaré. También te ayudará mi marido, si tú lo quieres. Me parece que eres un aprendiz fugitivo..., esto es todo. No tiene importancia. Has recibido malos tratos y has decidido acabar con todo. ¡Dios te bendiga, no pienso delatarte, hijo! Anda, sé buen chico y cuéntamelo todo.

Dije que sería inútil continuar fingiendo, que se lo confesaría todo, siempre que ella mantuviera su promesa. Entonces le conté que mis padres habían muerto, que la ley me había entregado a un granjero viejo y mezquino que vivía en el campo, a treinta millas del río, el cual me trataba tan mal, que no pude resistirlo más. Se marchó para pasar fuera dos días, de manera que aproveché la ocasión, robé algunas ropas de su hija y hui. Estuve tres días recorriendo las treinta millas. Viajé por las noches y de día me ocultaba y dormía. La bolsa con pan y carne que me llevé de la granja me había durado todo el camino. Dije también que confiaba en que mi tío Abner Moore cuidara de mí, y que por esta razón había llegado al pueblo de Goshen.

—¿Goshen, hijo? Esto no es Goshen. Este pueblo se llama San Petersburgo. Goshen está a una milla río arriba. ¿Quién te dijo que esto era Goshen?

—¡Oh!, pues un hombre que encontré hoy al amanecer, cuando iba a adentrarme en los bosques para dormir. Me dijo que doblara hacia la derecha cuando llegara al cruce de caminos y que encontraría Goshen cinco millas más adelante.

—Me figuro que estaría borracho. Te lo dijo al revés.

—Hablaba como si lo estuviera, pero ya no importa. Tengo que continuar mi camino. Llegaré a Goshen antes de que sea de día.

—Aguanta un minuto. Te prepararé algo de comer. Puedes necesitarlo.

Después de preparar la comida preguntó:
—Oye cuando una vaca está echada, ¿por qué parte se levanta primero? Contesta de prisa... No te detengas ni a pensarlo. ¿Por qué parte se levanta primero?
—Por atrás, señora.
—Bueno, ¿y un caballo?
—Por delante, señora.
—¿A qué lado de un árbol se cría más musgo?
—Por el norte.
—Si hay quince vacas paciendo en la ladera de una montaña, ¿cuántas de ellas comen con la cabeza hacia la misma dirección?
—Las quince, señora.
—Bien, me parece que sí has vivido en el campo. Pensé que tal vez tratabas de engañarme otra vez. ¿Cuál es tu nombre verdadero?
—George Peters, señora.
—Bueno, procura recordarlo, George. No vaya a olvidársete y me digas que te llamas Alexander antes de irte y luego me salgas con que te llamas George-Alexander. Y no te acerques a las mujeres llevando este viejo vestido de calicó. Aparentas una pobre niña bastante mediocre, aunque tal vez engañes a los hombres. Bendito seas, hijo, y cuando quieras enhebrar una aguja no sostengas quieto el hilo y le acerques la aguja; sostén quieta la aguja y acércale el hilo para pasarlo... Así es cómo suele hacerlo una mujer, pero el hombre lo hace al revés. Y cuando tires algo a una rata o a cualquier otra cosa ponte de puntillas y levanta la mano por encima de la cabeza tan torpemente como sepas, y no le des a la rata por lo menos por seis o siete pies de distancia. Lánzalo manteniendo el brazo rígido desde el hombro, como si ahí tuvieras un pivote que lo hace girar; no desde la muñeca y el codo, extendiendo el brazo a un lado, como lo hace un muchacho. Y recuerda que cuando una chica trata de sostener algo en el regazo separa las rodillas, no las junta, como has hecho tú para coger la barra de plomo. Me di cuenta de que eras un chico cuando enhebrabas la aguja y te preparé las otras pruebas para cerciorarme. Ahora corre al lado de tu tío, Sarah Mary Williams George Alexander Peters, y si te encuentras en apuros avisa a la señora Judith Loftus, que soy yo, y haré lo que pueda para sacarte del aprieto. Sigue el camino del río, y la próxima vez que debas hacer un largo recorrido a pie lleva zapatos y calcetines. El camino del río es rocoso y supongo que llegarás a Goshen con los pies destrozados.

Subí la ribera unas cincuenta yardas y luego desanduve lo andado y me dirigí al lugar donde estaba mi canoa, bastante más abajo de la casa. Salté dentro y me marché a toda prisa. Remonté la corriente lo suficiente para alcanzar el cabo de la isla y entonces empecé a cruzar. Me quité el sombrero porque ya no necesitaba la visera. Cuando me encontraba en el centro del río oí que el reloj empezaba a dar campanadas. Me paré a escuchar. El sonido llegaba débilmente por encima del agua, pero con claridad... Eran las once. Cuando llegué a la isla no me detuve a tomar aliento, aunque estaba desfallecido, sino que me interné entre los árboles, donde antes solía tener mi campamento, y encendí una buena hoguera en un sitio seco.

Luego salté dentro de la canoa y me dirigí hacia nuestro escondite, una milla y media más abajo, tan rápidamente como pude. Desembarqué y atravesé el bosque, subí la colina y entré en la caverna. Allí estaba Jim profundamente dormido en el suelo. Le desperté diciéndole:

—¡Levántate y muévete, Jim! ¡No hay un minuto que perder! ¡Vienen por nosotros!

Jim no hizo preguntas, ni siquiera dijo una palabra. Pero el modo como trabajó durante la siguiente media hora demostró lo asustado que estaba. Entonces todo cuanto teníamos en el mundo estaba en nuestra balsa, la cual estaba lista para desatracar de la caleta de sauces donde estaba escondida. Primero apagamos el fuego de la caverna y después no sacamos fuera ni una vela.

Me alejé un poco de la orilla en la canoa para echar un vistazo, pero, si había un bote, no pude verlo, porque las estrellas y las sombras no sirven para ver bien. Luego sacamos la balsa y nos deslizamos bajo la sombra, pasando junto a la isla, mortalmente quietos, sin pronunciar una palabra.

12

Faltaría poco para la una cuando por fin estuvimos más abajo de la isla, y la balsa avanzaba muy despacio. Si se aproximaba alguna embarcación, cogeríamos la canoa y nos dirigiríamos a la orilla de Illinois. Menos mal que no se acercó ningún bote, porque ni se nos ocurrió meter la escopeta dentro de la canoa, ni una caña de pescar, ni nada de comer. Estábamos demasiado preocupados para pensar en tantas cosas. No fue muy acertada la idea de ponerlo todo en la balsa.

Si los hombres visitaron la isla, supongo que encontraron la hoguera que yo encendí y esperaron toda la noche a que apareciese Jim. En todo caso, estuvieron lejos de nosotros y, si mi treta del fuego no los engañó, la culpa no fue mía. Lo hice con tanta mala intención como pude.

Cuando empezó a amanecer, amarramos a un saliente de remolque en un marcado meandro del lado de Illinois, cortamos ramas de chopos de Virginia con la hacheta y con ellas cubrimos la balsa; para darle el aspecto de un corrimiento de tierra en el banco que había allí. Un saliente de remolque es un banco de arena que tiene chopos tan compactos como los dientes de una traílla.

Teníamos montañas en la orilla de Missouri y densa arboleda en la de Illinois, y en aquel lugar el canal estaba en la orilla de Missouri, de modo que no temíamos que nadie nos encontrara. Pasamos allí el día, tumbados, observando las balsas y los vapores que navegaban por la orilla del Missouri, y los vapores que remontaban el río, luchando con él en el centro. Hablé a Jim de mi aventura con aquella mujer, y Jim dijo que era muy lista y que, si ella nos seguía la pista, seguro que no se sentaba a montar guardia junto a la hoguera del campamento... No, señor, traería un perro.

—Bueno —dije yo entonces—, ¿y si ella dijo a su marido que llevara un perro?

Jim apostó a que ella no lo pensó cuando los hombres se marcharon, y creía que deberían subir al pueblo en busca de un perro, que perderían mucho tiempo, pues de lo contrario nosotros no nos encontraríamos ahora a dieciséis o diecisiete millas más abajo del pueblo... No, señor, estaríamos en el pueblo de nuevo. Dije que no importaba el motivo por el que no nos capturasen, siempre que no lo lograran.

Cuando empezó a anochecer, asomamos las cabezas entre la densa barrera de los chopos y miramos en todas direcciones. No se veía nada. Jim cogió algunas tablas superiores de la balsa e improvisó una choza en la que guarecernos de la lluvia y el mal tiempo, así como para mantener las cosas secas. También hizo un suelo para la choza, levantándolo algo más de un pie por encima del nivel de la balsa, a fin de que las mantas y todo lo demás no fuera alcanzado por el oleaje de los vapores. Al centro pusimos una capa de tierra de unas cinco o seis pulgadas de grueso, enmarcada, a fin de que no se moviera de sitio. Serviría para encender una hoguera cuando el tiempo empeorase o hiciera frío. También hicimos un remo extra, porque uno de los otros podía romperse contra un tronco flotante o algo parecido. Fijamos un palo corto y ahorquillado para colgar la vieja linterna, porque siempre debíamos encenderla cuando viésemos bajar un vapor por el río, con objeto de impedir que nos arrollase, pero no haría falta encender la linterna para los vapores que remontaban el río, a menos que estuviéramos en lo que llaman un cruce, pues el río aún estaba bastante crecido y sus orillas, muy bajas, seguían inundadas, de manera que los vapores que remontaban el río no siempre seguían el canal, sino que preferían el agua fácil.

Esa segunda noche caminamos unas siete u ocho horas, con una corriente que hacía más de cuatro millas por hora. Pescamos, hablamos y nos bañamos de vez en cuando para no dormirnos. Daba una sensación de solemnidad dejarnos llevar a la deriva río abajo, por sus aguas quietas, tumbados boca arriba, contemplando las estrellas, y ni siquiera deseábamos hablar en voz alta, y no reíamos, sino que soltábamos una risita entre dientes. En conjunto tuvimos buen tiempo y no nos ocurrió nada aquella noche, ni la siguiente, ni la tercera.

Todas las noches pasábamos frente a algunos pueblos. Los había situados en lo alto de negras colinas, como un radiante lecho de luces, sin que pudiera verse una sola casa. La quinta noche pasamos por delante de St. Louis y pareció como si el mundo entero estuviera iluminado. Solían decir en San Petersburgo que había veinte o treinta mil personas en St. Louis, pero nunca lo creí hasta ver aquel maravilloso derroche de luces a las dos de aquella noche tranquila. No se oía ningún ruido, todo el mundo dormía.

Cada noche solía bajar a tierra, alrededor de las diez, para comprar diez o quince centavos de comida, tocino o cualquier otra cosa, en un pueblecito. A veces escamoteaba un pollo que no dormía cómodamente en su gallinero y me lo llevaba. Papá decía siempre que hay que coger un pollo cuando se tiene la oportunidad, porque, si uno no lo quiere, fácilmente encontrará alguien que lo quiera, y una buena acción jamás se olvida. Nunca vi que papá no quisiera el pollo, pero en todo caso eso era lo que él decía.

Por las mañanas, antes de amanecer, me deslizaba dentro de sembrados y me llevaba una sandía, un melón, una calabaza o un poco de maíz nuevo y cosas por el estilo. Papá decía siempre que no era malo llevarse cosas prestadas si uno tenía el propósito de pagarlas algún día, pero la viuda decía que eso no era más que una expresión suave de la palabra «robar», y que ninguna persona honrada debía hacerlo. Jim dijo que le parecía que tanto la viuda como papá llevaban la mitad de razón, de modo que lo mejor para nosotros sería elegir dos o tres cosas de la lista y decir que nunca las tomaríamos prestadas... Luego, suponía que no había nada malo en tomar prestado el resto.

Así que lo hablamos durante toda una noche, mientras flotábamos río abajo, tratando de decidir si nos desprendíamos de las sandías o de los melones o de qué. Pero hacia el amanecer llegamos a un acuerdo satisfactorio. Nos desprenderíamos de las manzanas silvestres y de los nísperos. Antes no estábamos tranquilos, pero ya estaba resuelto el problema. Me alegré de la decisión tomada, porque las manzanas silvestres nunca son del todo buenas y los nísperos tardarían aún dos o tres meses en madurar.

Cazamos de vez en cuando algún pájaro que se despertaba demasiado temprano por la mañana o se acostaba bastante tarde por la noche. En conjunto, vivíamos estupendamente.

A la quinta noche, más abajo de St. Louis, tuvimos una gran tormenta después de medianoche, con fuertes truenos y relámpagos, y la lluvia cayó como si fuera cortina de agua sólida. Nos quedamos en la choza y dejamos que la balsa se cuidara de sí misma. Con el resplandor de los relámpagos pudimos ver delante un gran río, flanqueado por altos acantilados rocosos. Dije:

—¡Caramba, mira allá, Jim!

Era un vapor que se había estrellado contra una roca. La corriente nos llevaba hacia él. El relámpago nos lo dejó ver claramente. Estaba ladeado, con parte de su cubierta superior asomando por encima del agua, y se podía ver nítidamente la chimenea y una silla junto a la gran campana, con un chambergo colgado del respaldo, cuando lo iluminaban los relámpagos.

Bueno, quizá porque era una noche de tempestad y todo tenía tanto aire misterioso, sentí exactamente lo que cualquier muchacho al ver el barco naufragado allí, en el centro del río, sombrío y solitario. Deseé subir a bordo y recorrerlo para ver qué había dentro. De modo que dije:

—Desembarquemos en él, Jim.

Pero Jim se oponía al principio:

—No quiero hacer el tonto visitando un naufragio. Las cosas nos van bien y es mejor dejarlas como están, según dice el libro divino. Apuesto a que hay un guardián a bordo.

—¡Que va a haber una guardián, hombre de Dios! —repliqué yo—. No hay nada que vigilar, salvo el sombrero y el timón. ¿Crees que alguien se expondría por un sombrero y un timón en una noche como ésta, cuando puede que se parta por la mitad en cualquier momento y sea arrastrado río abajo? —Jim no pudo oponer nada a esto, de modo que no lo intentó—. Además —continué—, podíamos tomar prestado lo que valiera la pena de la cabina del capitán. Te apuesto a que hay cigarros... de los de cinco centavos cada uno. Los capitanes de los vapores siempre son ricos y ganan seis dólares al mes, y les importa un comino lo que cuesta algo que les interesa, ¿entiendes? Métete una vela en el bolsillo. No estaré tranquilo, Jim, hasta que le echemos un vistazo por dentro. ¿Crees que Tom Sawyer despreciaría semejante bicoca? ¡Ni hablar! Lo llamaría una aventura, eso es... Y desembarcaría en este barco naufragado, aunque fuera lo último que hiciera con vida. ¡Y no le echaría imaginación a la aventura! ¡Te apuesto a que te parecería estar viendo a Cristóbal Colón descubriendo el Nuevo Mundo! ¡Ojalá Tom Sawyer estuviera aquí!

Jim gruñó, pero al fin accedió. Dijo que no habláramos más de lo preciso y en voz baja. Al resplandor de los relámpagos vimos de nuevo el buque naufragado, nos agarramos a la cabria de estribor y amarramos allí mismo.

La cubierta estaba alta en ese punto. Nos escurrimos hacia abajo por la pendiente hacia babor, en la oscuridad, tanteando suavemente el suelo con los pies y alargando las manos para esquivar las retenidas, porque estaba tan a oscuras que no podíamos ni verlas. Pronto tropezamos con el extremo proel de la claraboya y nos encaramamos a ella. El paso siguiente nos llevó ante la puerta de la cabina del capitán, que estaba abierta, y, ¡diantres!, abajo, en la cámara, vimos una luz. ¡Y en el mismo momento nos pareció oír unas voces quedas allá!

Jim dijo en un susurro que se encontraba muy mal y que lo siguiera. Respondí que estaba de acuerdo y me encaminaba ya hacia la balsa, cuando oí una voz lastimera que decía:

—¡Por favor, muchachos! ¡Juro que no lo diré a nadie!

Otro respondió en voz bastante alta:

—¡Mientes, Jim Turner! Siempre pediste más y lo conseguiste porque nos amenazabas con delatarnos si no te lo dábamos. Pero esta vez has ido demasiado lejos. Eres el perro más canalla y traidor del país.

Jim se había alejado ya hacia la bolsa. Yo me sentía devorado por la curiosidad y me dije que Tom Sawyer no se volvería atrás ahora, ni yo tampoco. Vería qué pasaba allí. De modo que me agaché y a cuatro patas avancé por el pequeño corredor, a oscuras, hasta que no quedaba más que un camarote entre el salón y yo. Allí vi a un hombre tendido en el suelo, atado de pies y manos, y a dos hombres de pie junto a él, uno de los cuales sostenía una pequeña linterna en la mano y el otro encañonaba con una pistola al hombre tendido en el suelo, al tiempo que le decía:

—¡Me gustaría hacerlo! ¡Y tendría que hacerlo, granuja inmundo!

El hombre que yacía en el suelo se encogía diciendo:

—¡Por favor, no lo hagas, Bill...! ¡No diré nada!

Y cada vez que decía eso el hombre de la linterna se reía diciendo:

—¡Pues claro que no lo dirás! ¡Apuesto a que nunca has dicho una verdad! —Y luego añadió—: ¡Cómo suplica! Y, sin embargo, si no llegamos a adelantarnos atándolo, nos habría matado a los dos. ¿Y por qué? Por nada. Simplemente porque reclamábamos nuestros derechos..., por eso nada más. Pero creo que no volverás a amenazar a nadie, Jim Turner. ¡Guárdate la pistola, Bill!

Bill dijo:

—No lo haré, Jake Packard. Prefiero matarlo... ¿No mató él al viejo Hatfeld de la misma manera? ¿No se lo merece?

—Pero es que no quiero matarlo y tengo buenas razones.

—¡Dios te bendiga por estas palabras, Jake Packard! ¡No las olvidaré mientras viva! —dijo el hombre del suelo, casi gimoteando.

Packard no le hizo el menor caso, sino que colgó su linterna de un clavo y se encaminó hacia donde estaba yo, a oscuras, e indicó a Bill con un ademán que le siguiera. Retrocedí a gatas tan de prisa como pude unas dos yardas, pero, al estar tan ladeado el barco, no adelanté mucho, de modo que para evitar que tropezaran conmigo y me sorprendieran entré en un camarote de la parte de arriba. El hombre avanzó a tientas y, cuando llegó a mi camarote, dijo:

—Ven... Entra aquí.

Y entró seguido de Bill. Pero antes de que entraran yo estaba ya en la litera, acurrucado y arrenpetido de haber subido a bordo. Allí estaban ellos, con las manos sobre el borde de la litera, hablando. No pude verlos, pero adivinaba dónde estaban por el olor de whisky que echaban. Me alegré de no beber whisky, aunque esto no hubiera cambiado las cosas, porque no habrían podido descubrirme, ya que ni siquiera respiraba. Estaba demasiado asustado. Además, uno no podía respirar escuchando semejante conversación. Hablaban en tono bajo y grave. Bill quería matar a Turner. Dijo:

—Ha dicho que me delatará y lo hará. Si le diéramos nuestras dos partes ahora, las cosas seguirán igual después de la pelea y de cómo le hemos tratado. Tan seguro como que estás vivo, que él nos denunciará. Ahora escúchame: Creo que debemos eliminarlo.

—Yo también —dijo Packard tranquilamente.

—¡Maldición, empezaba a sospechar lo contrario! Bueno, pues todo arreglado. Vamos a eliminarlo.

—Aguarda un minuto. No me has dejado hablar. Escúchame tú a mí. No está mal pegarle un tiro, pero hay medios más discretos, ya que hay que hacer el trabajo. Lo que quiero decir es esto: Es de locos andar solicitando una soga para el cuello cuando uno puede conseguir su propósito de otra manera que no acarrea riesgos. ¿Tú qué crees?

—Que tienes toda la razón. Pero ¿cómo lo solucionarás esta vez?

—Mi idea es ésta, verás: Registraremos todo el barco cogiendo las cosas que se nos pasaron por alto en los camarotes, bajaremos a la orilla y esconderemos el botín. Entonces esperaremos. Apuesto a que no pasan ni dos horas antes de que este barco se desmorone y sea arrastrado río abajo. ¿Lo entiendes? El se ahogará y nadie tendrá la culpa más que él. Me figuro que es mucho mejor esto que matarlo. No soy partidario de matar a un hombre mientras pueda evitarse. No es sensato ni moral. ¿Me equivoco?

—No... Creo que no. Pero supón que el barco no se desmorona ni se lo lleva el río.

—Bueno, esperemos dos horas y veamos qué pasa, ¿no?

—De acuerdo, vamos.

Nada más que se alejaron, yo me largué, envuelto en sudor frío, hacia la proa. Aquello estaba oscuro como boca de lobo. Llamé con un susurro áspero:

—¡Jim!

El me contestó con un gemido a mi lado. Yo le dije entonces:

—¡Jim, no podemos perder tiempo gimoteando! Es una banda de asesinos y, si no encontramos su bote y lo dejamos a la deriva en el río para que esos tipos no puedan huir, uno de ellos las va a pasar negras. Pero, si encontramos su bote, podemos meterlos a todos en el aprieto... y el sheriff se encargará de ellos. ¡Pronto... date prisa! Yo miraré por el lado de babor y tú por el de estribor. Empiezas por la balsa y...

—¡Ay, Señor, mi Señor! ¿La balsa? ¡No hay ninguna balsa! Se han soltado las amarras y ha desaparecido... ¡Y nosotros nos hemos quedado aquí!

13

Bueno, el caso es que me quedé sin aliento y casi me desmayé. ¡Acorralados en el barco con semejante banda! Pero no había tiempo que perder para tener flaquezas sentimentales. Teníamos que encontrar el bote ahora... para nosotros. De modo que, temblando, bajamos por el lado de estribor, y fue una maniobra tan lenta, que se nos antojó que pasaba una semana antes de que alcanzáramos la popa. Ni rastro de un bote. Jim dijo que no podía seguir adelante; estaba tan asustado, que le faltaban las fuerzas. Pero yo le dije que, si nos quedábamos a bordo del barco naufragado, estábamos perdidos sin remedio. De modo que seguimos andando al acecho. Buscamos la popa del camarote y lo encontramos. Luego nos dirigimos hacia la proa por la claraboya, colgándonos de persiana en persiana, porque el saliente de la claraboya estaba en el agua. Cuando llegamos cerca de la puerta del salón, ¡allí estaba el esquife! Apenas podía verlo.

Me sentí aliviado. Un segundo más y estaría a bordo del esquife, pero justamente entonces se abrió la puerta. Uno de los hombres asomó la cabeza a unos dos pies de mí, y pensé que estaba ya perdido, pero la metió nuevamente dentro, diciendo:

—¡Aparta de una vez esa maldita linterna, Bill!

Arrojó una bolsa o algo en el bote y luego saltó dentro y se sentó. Era Packard. Acto seguido salió Bill y se reunió con él. Packard dijo en voz baja:

—Preparados... ¡Vámonos!

Me fue difícil seguir colgándome de las persianas, por lo débil que me sentí. Pero Bill dijo:

—Aguarda, ¿le has registrado?

—No. ¿No lo hiciste tú?

—No. Así que tiene aún su parte del dinero.

—Bueno, vamos entonces... Es inútil llevarnos el botín y dejar el dinero.

—Oye... ¿No sospechará lo que tramamos?

—Tal vez no. Pero, de todo modos, hemos de quitárselo. Vamos.

Dejaron el bote y volvieron a entrar.

La puerta se cerró de golpe porque estaba en el costado carenado: medio segundo después yo estaba en el bote y Jim me seguía a tropezones. ¡Corté la cuerda con mi cuchillo y nos largamos!

No tocamos un remo, no hablamos ni en susurros, apenas respiramos. Nos dejábamos llevar por la rápida corriente, en un silencio de muerte, por delante de la punta del tambor de rueda y de la popa. Un par de segundos más tarde nos encontrábamos a cien yardas más abajo del barco naufragado, que las tinieblas engulleron sin dejar rastro. Estábamos salvados y lo sabíamos.

Cuando estábamos a tres o cuatrocientas yardas, vimos aparecer la linterna como una chispita en la puerta del camarote, por un segundo, y comprendimos que los granujas habían descubierto la desaparición del bote y se daban cuenta de que se encontraban en el mismo apuro que Jim Turner.

Entonces Jim empuñó los remos y emprendimos la búsqueda de nuestra balsa. Fue entonces cuando empecé a preocuparme por los hombres. Creo que antes me faltó tiempo. Empecé a pensar en lo espantoso que debía ser, incluso para unos asesinos, encontrarse en semejante aprieto. Me dije que quién podía asegurar que no me convirtiera en un asesino, y que, entonces, ¿me gustaría que me hicieran aquello? Por lo tanto, dije a Jim:

—En cuanto veamos una luz, desembarcaremos cien yardas más arriba o más abajo de ella, en un lugar donde haya un buen escondrijo para ti y el esquife, y entonces idearé una historia fantástica para que alguien vaya en busca de esa banda y los saque del apuro, para que les cuelguen cuando les llegue el momento.

Pero la idea fue un fracaso. Pronto se desencadenó otra tormenta, y esta vez peor que nunca. Llovía intensamente y no se veía ni una luz.

«Todo el mundo debe estar acostado», pensé. Descendimos por el río en busca de luces y de nuestra balsa. Al cabo de largo rato cesó de llover, pero quedaron las nubes, los relámpagos continuaron gimiendo con furia y a poco uno de sus destellos nos reveló un objeto negro que flotaba ante nosotros y hacia el cual nos dirigimos.

Era la balsa, y nos alegró una enormidad subir a bordo de ella nuevamente. Vimos entonces una luz a la derecha, en la orilla. Dije que iría hacia allí. El esquife estaba medio lleno del botín que había robado la banda del buque. Lo trasladamos a la balsa e indiqué a Jim que siguiera flotando y encendiera una luz cuando calculase que llevaba recorridas dos millas, sin apagarla hasta mi regreso. Luego empuñé los remos y me dirigí hacia la luz. Mientras descendía vi tres o cuatro luces más en la ladera de la colina. Era un pueblo. Rodeé la luz de la orilla, levanté los remos y seguí flotando. Al pasar vi una linterna colgada del asta de la bandera de un vapor de doble quilla. Miré alrededor en busca del vigilante y preguntándose dónde dormiría. A poco le encontré durmiendo en las bitas de proa, con la cabeza entre las rodillas. Le sacudí dos o tres veces por el hombro y empecé a llorar.

Se despertó bastante sobresaltado, pero, cuando vio que sólo era yo, bostezó a gusto, estiró los brazos y dijo:

—Hola, ¿qué pasa? No llores, peque. ¿Qué te ocurre?

Yo dije:

—Papá, mamá, mi hermanita y...

Entonces me deshice en lágrimas. El dijo:

—Oh, vamos, vamos; no lo tomes de esta manera. Todos tenemos problemas y éste tuyo se solucionará. ¿Qué ocurre?

—Ellos... ellos... ¿Es usted el vigilante del barco?

—Sí —contestó, bastante satisfecho—. Soy el capitán, el dueño, el piloto, el vigilante y el primer marinero, y a veces la carga y el pasaje. No soy tan rico como el viejo Jim Hornsback y no puedo ser tan generoso y bueno para con Tom, Dick y Harry como lo es él, y gastar el dinero que se gasta él, pero le he dicho más de una vez que no me cambiaría por él, porque es lo que yo le digo: la vida de marinero es la mía, y que me cuelguen si vivo alguna vez a dos millas del pueblo, donde nunca pasa nada; no lo haría por toda su pasta y mucha más que me dieran. Es lo que yo digo...

Le interrumpí para decir:

—Están en un espantoso atolladero y...

—¿Quiénes?

—Pues papá, mamá y mi hermanita y la señorita Hooker; y si usted cogiera su vapor y fuera allí...

—¿A dónde? ¿Dónde están?

—En el barco naufragado.

—¿Qué barco?

—Pues no hay más que uno.

—¿Cómo? ¿Hablas del «Walter Scott»?

—Sí.

—¡Cielo Santo! Por el amor de Dios, ¿qué hacen allí?

—Pues no fueron a propósito.

—¡Apuesto a que...! Pero, Dios bendito, ¡están perdidos si no se va de prisa! Y ¿cómo diantres se metieron en ese atolladero?

—Muy fácil. La señorita Hooker estuvo de visita en el pueblo...

—Sí, en Boots's Landing... Continúa.

—Estuvo allí y, al anochecer, salió con su ama negra en la barcaza para pasar la noche en casa de su amiga, la señorita no sé cómo se llama. Perdieron el remo de gobierno, dieron vueltas y entonces flotaron río abajo, de popa, unas dos millas, y luego quedó montada sobre el barco naufragado, y el barquero, la negra y los caballos se perdieron, pero la señorita Hooker logró agarrarse y subir a bordo del barco naufragado. Bueno, pues una hora después de anochecer llegamos nosotros en nuestra chalupa mercante y estaba tan oscuro, que no vimos el barco hasta que estuvimos encima, y también nosotros quedamos montados sobre él; nos salvamos todos, menos Bill Whipple... ¡Oh, qué buena persona era! Casi preferiría haber sido yo...

—¡Válgame Dios; es la cosa más extraordinaria que he oído en mi vida! Y entonces, ¿qué hicisteis?

—Pues gritamos continuamente, pero el río es tan ancho allí, que nadie nos oyó. Papá dijo que alguien debía bajar a tierra en busca de auxilio. Yo era el único que sabía nadar, de modo que me ofrecí, y la señorita Hooker dijo que si no encontraba pronto ayuda viniese aquí en busca de su tío, que él lo solucionaría. Toqué tierra a una milla más abajo de aquí y desde entonces he ido de un lado a otro tratando de lograr que la gente hiciera algo por mí, pero todos decían: «¿Cómo, en semejante noche y con esa corriente? Es una locura. Ve al vapor.» Ahora, si usted va y...

—¡Diablos, me gustaría ir, y no sé aún si iré! Pero ¿quién pagará después el gasto? ¿Crees tú que tu papá...?

—¡Oh, claro está que sí! La señorita Hooker recalcó que su tío Hornsback...

—¡Rayos y truenos! ¿Es él su tío? Oye, ve corriendo hacia aquella luz de allá y dobla hacia el oeste. Un cuarto de milla más adelante llegarás a la taberna. Diles que te lleven a casa de Jim Hornsback y él pagará la cuenta. Y no te entretengas, porque él querrá conocer la noticia. Dile que habré salvado a su sobrina antes de que él llegue al pueblo. Anda, echa a correr. Voy ahí, a la vuelta de la esquina, a despertar a mi maquinista.

Corrí hacia la luz, pero en cuanto él desapareció tras la esquina volví sobre mis pasos, salté dentro de mi esquife, achiqué el agua y luego remé orilla arriba, en el agua mansa, unas seiscientas yardas y me escondí entre algunas barcas, porque no estaría tranquilo hasta ver salir el vapor. Pero en conjunto me sentía bastante satisfecho de la manera en que ayudaba a aquella banda, pues no lo harían muchos en mi lugar. Deseé que la viuda lo supiera. Pensé que ella estaría orgullosa de mí por ayudar a aquellos bribones, porque los granujas y los desgraciados son los que más interesan a la viuda y a las buenas personas.

¡Bueno, a poco bajó a la deriva el barco naufragado, sombrío y silencioso! Me recorrió una especie de temblor frío y luego remé hacia él. Estaba muy hundido y pronto me di cuenta de que había pocas posibilidades de que dentro hubiera alguien con vida. Di la vuelta alrededor y lancé unos gritos, pero no hubo respuesta. Había un silencio de muerte. Me sentí algo apenado por la banda, pero no mucho, porque pensé que, si ellos podían soportarlo, yo también podría.

Entonces apareció el vapor, de modo que me dirigí hacia el centro del río diagonalmente. Cuando calculé que estaba fuera de su vista, levanté los remos y miré atrás. Vi que el vapor husmeaba el barco naufragado en busca de los restos de la señorita Hooker, porque el capitán sabía que tío Hornsback querría tenerlos. Luego el vapor abandonó bastante pronto la búsqueda y regresó a la orilla. Volví a remar y me lancé a toda velocidad río abajo.

Se me antojó que pasaba larguísimo rato antes de aparecer la luz de Jim. Cuando la vi, parecía hallarse a mil millas de distancia. Cuando la alcancé, el cielo empezaba a volverse grisáceo por el este, de manera que nos dirigimos hacia una isla, ocultamos la balsa, hundimos el esquife, nos tumbamos y quedamos dormidos como muertos.

14

Poco después de levantarnos repasamos las cosas que la banda robó del barco naufragado y hallamos botas, mantas, ropas y toda suerte de objetos, así como un montón de libros, un catalejo y tres cajas de cigarros. Ninguno de los dos había sido tan rico en toda su vida. Los cigarros eran de primera. Estuvimos toda la tarde tumbados en los bosques, hablando. Yo hojeé los libros y lo pasamos bastante bien. Conté a Jim todo lo courrido dentro del barco y en el vapor. Dije que cosas así eran aventureras, pero él dijo que estaba harto de aventuras. Añadió que, cuando yo entré en el camarote y él se volvió a gatas hacia la balsa y descubrió que había desaparecido, estuvo a punto de morirse del susto, porque pensó que él estaba perdido de todos modos, porque, si no se salvaba, moriría ahogado, y, si se salvaba, el que le salvara le devolvería a su ama por la recompensa ofrecida, y entonces la señorita Watson le vendería a los del sur sin dudarlo. Bueno, Jim tenía razón; casi siempre la tenía. Para ser negro, poseía un cerebro extraordinario.

Leí muchas cosas a Jim respecto a los reyes, duques, condes y gente así. Se enteró de los suntuosos ropajes que llevaban, de su gran estilo, de que entre sí se llamaban Majestad y Su Señoría y cosas así, en lugar de señor. Los ojos se le desorbitaban a Jim; estaba muy interesado, y dijo:

—No sabía que hubiera tantos. No había oído hablar de ninguno de ellos; es decir, solamente del rey Salomón, a menos que llames reyes a los de la baraja. ¿Cuánto gana un rey?

—¿Cuánto gana? —exclamé yo—. ¡Pues mil dólares al mes, si quiere! Los reyes tienen todo lo que desean, es todo suyo.

—¿Verdad que es estupendo? Y ¿qué tienen que hacer, Huck?

—¡No hacen nada! ¡Qué cosas dices...! No hacen más que estar sentados...

—¡No! ¿De veras?

—¡Pues claro! Están siempre sentados menos cuando hay guerra. Entonces tienen que ir a la guerra, pero normalmente están ganduleando por ahí, o van de caza con el halcón... con el halcón y... ¡Chisst! ¿Has oído ese ruido?

Atisbamos, pero no había nada anormal, simplemente el rumor de la rueda del vapor que descendía doblando el cabo. Volvimos al sitio.

—Sí —continué yo—, y en otras ocasiones, cuando se aburre, alborota con el parlamento y, si no hacen todos lo que él quiere, les corta la cabeza. Pero lo que suelen hacer los reyes es rondar el harén.

—Rondar... ¿el qué?

—El harén.

—¿Qué es eso?

—El sitio donde guardan a sus mujeres. ¿No has oído hablar del harén? Salomón tenía uno y tenía aproximadamente un millón de esposas.

—¡Ah, sí, es verdad! Yo... lo había olvidado. Me figuro que un harén es algo así como una pensión. Lo más seguro es que tengan muchas complicaciones en el cuarto de los niños. Y supongo que las mujeres discutirán muchísimo, y así el alboroto se hace mayor. Y, sin embargo, dicen que Salomón fue el hombre más sabio que ha vivido jamás. No me convence. Porque ¿cómo se explica que un hombre tan sabio quisiera vivir en medio de semejante batahola? No... desde luego, no es posible. Un hombre sabio se construiría una fábrica de calderos y, cuando quisiera descansar, no tendría más que cerrar la fábrica y en paz.

—Bueno, pues fue el hombre más sabio, porque la viuda me lo dijo.

—No sé qué te diría la viuda, pero él no fue un sabio. Casi todas las cosas que hizo tenían truco. ¿Sabes lo del crío que iba a partir en dos?

—Sí, me lo contó la viuda.

—¡Pues entonces...! ¿No fue una idea disparatada? No tienes más que pensarlo un minuto. Ahí tienes ese tocón... Digamos que es una de las mujeres. Ahí estás tú... Eres la otra mujer. Yo soy Salomón y este billete de dólar es el niño. Las dos lo reclamáis. ¿Qué hago yo? ¿Empiezo a preguntar entre los vecinos para averiguar a cuál de las dos pertenece el billete y se lo doy a la mujer que es la dueña, sano y salvo, como haría una persona con la cabeza bien sentada? No... Voy y parto el billete en dos; te doy la mitad a ti y la otra mitad a la otra mujer. Así es cómo pensaba hacerlo Salomón con el niño. Y ahora te pregunto yo: ¿De qué sirve medio billete? No puedes comprar nada con él. ¿Y de qué sirve medio crío? No daría ni tanto así por un millón de mitades de niños.

—¡Diantre, Jim; a ti se te ha escapado el verdadero sentido, hombre!

—¿A mí? Anda, chico, no me hables a mí de sentidos. Yo veo el sentido cuando lo hay, y ese asunto no lo tiene. No se disputaban medio crío, se disputaban al niño entero. Y el hombre que piensa que arreglará las cosas dando medio crío cuando uno lo quiere entero es que no sabe ni entrar en casa cuando llueve. No me hables de ese Salomón, Huck, le conozco por la espalda.

—¡Te digo que no entiendes el sentido!

—¡Y dale con el sentido! Yo sé lo que sé. Te repito que el verdadero sentido es más profundo. ¡Vaya si lo es! Viene de cómo se educó Salomón. Toma a un hombre que sólo tiene uno o dos niños. ¿Crees que ese hombre malgastará críos? No, señor, no los malgastará; no puede darse este gusto. El sabe valorarlos. Pero toma a un hombre que tiene unos cinco millones de críos correteando por su casa, y la cosa cambia. A él le resulta tan fácil partir a un niño en dos como si fuera un gato. Tiene muchos más. Niño más, niño menos, ¿qué podía importarle a Salomón? ¡Qué condenado!

Jamás había visto un negro igual. Si se le metía una idea en la cabeza, no había manera de sacársela. De los negros que conocía, era el que más inquina tenía a Salomón. Así que seguí hablando de otros reyes y dejé a Salomón a un lado. Le hablé de Luis XVI, al que le cortaron la cabeza en Francia mucho tiempo atrás, y de su hijo, el Delfín, que debía ser rey, pero al que se lo llevaron para encerrarlo en la cárcel, donde al cabo de algún tiempo se murió.

—¡Pobre chiquillo!

—Pero algunos dicen que se escapó y vino a América.

—¡Es estupendo! Pero se encontraría bastante solo... Aquí no hay reyes, ¿verdad, Huck?

—No.

—Entonces no puede conseguir un empleo. ¿Qué va a hacer?

—Pues no lo sé. Los hay que se meten a policías y otros que enseñan a la gente a hablar en francés.

—Oye, Huck, ¿los franceses no hablan del mismo modo que nosotros?

—No, Jim. No entenderías ni una palabra de lo que dicen... ni una palabra.

—¡Vaya, esto sí que me revienta! ¿Cómo es esto?

—Yo no lo sé, pero así es. Aprendí algo de su jerga en un libro. Supón que se acerca un hombre y te dice: *Parlé-vú fransé*... ¿Qué pensarías?

—No pensaría nada. Le daría un porrazo en la cabeza. Es decir, si no fuera blanco. No consentiría a ningún negro que me llamara eso.

—¡Caray, que no te llama nada! Solamente te pregunta si sabes hablar francés.

—Ah, pues entonces, ¿por qué no lo dice?

Vi a un hombre tendido en el suelo, atado de pies y manos, y a dos hombres de pie junto a él, uno... sostenía una linterna... y el otro una pistola... (pág. 57).

—¡Si te lo dice! Es la manera francesa de decírtelo.
—¡Qué cosa tan ridícula! Mira, no quiero hablar más de eso. No tiene sentido.
—Oye, Jim. ¿Un gato habla como nosotros?
—No, un gato no.
—Bueno, ¿y una vaca?
—No, una vaca tampoco.
—¿Habla un gato como una vaca o una vaca como un gato?
—No.
—Es natural y correcto que cada uno hable de manera distinta, ¿verdad?
—¡Claro!
—¿Y no es natural y correcto que un francés hable de distinta manera que nosotros? Anda, contéstame a eso.
—¿Un gato es un hombre?
—No.
—Entonces, no tiene sentido que un gato hable como un hombre. ¿Una vaca es un hombre? ¿O bien es un gato?
—Ni una cosa ni otra.
—Bien, pues la vaca no tiene por qué hablar como uno o el otro. ¿Un francés es un hombre?
—Sí.
—¡Ah, bien! ¿Por qué, pues, el muy condenado no habla como un hombre? ¡Contéstame tú a eso!
Vi que era inútil seguir hablando... Es imposible enseñar a discutir a un negro. Así que me di por vencido.

15

Calculamos que al cabo de otras tres noches llegaríamos a Cairo, al fondo del Illinois, donde desemboca el río Ohio, y eso era lo que buscábamos. Venderíamos la balsa, embarcaríamos en un vapor y remontaríamos el Ohio entre los Estados libres para poner término a nuestras tribulaciones.

Bueno, la segunda noche apareció la niebla y nos dirigimos a un asidero de remolque, pues nada sacaríamos con seguir navegando con la niebla. Cuando avancé remando en la canoa, con el cable, para amarrar, no encontré más que árboles muy pequeños. Até el cable alrededor de uno que había en el borde de la ribera cortada, pero había una corriente muy fuerte, y la balsa descendió tan rápidamente, que arrancó el árbol de raíz llevándoselo río abajo. Vi que se espesaba la niebla y me asusté tanto que casi pasé medio minuto sin moverme... y entonces la balsa había desaparecido ya de mi vista. Era imposible ver más allá de veinte yardas. Salté dentro de la canoa, corrí a popa, cogí la pértiga y empecé a remar. Pero no se movía. Con tantas prisas, no la había desamarrado. Me levanté e intenté desatar el cable, pero con la excitación las manos me temblaban y apenas pude hacer nada.

En cuanto me puse en marcha perseguí la balsa acalorado, inquieto. Al principio todo fue bien, pero el asidero de remolque no tenía sesenta yardas de largo, y en cuanto pasé por delante me adentré a toda velocidad en la sólida niebla blanca y quedé tan desorientado de la dirección que llevaba como lo estaría un muerto.

Pensé, «No servirá de nada que reme; antes de que me dé cuenta chocaré contra la ribera, un asidero de remolque o cualquier otra cosa. Tengo que permanecer quieto y seguir flotando; sin embargo, resulta una verdadera tortura mantener las manos quietas en semejante momento». Grité y escuché. Allá abajo, lejos, oí un grito que levantó mis ánimos. Continué avanzando, atento el oído por si volvía a oír el grito. La segunda vez me di cuenta de que, en lugar de acercarme al lugar de donde procedía, me alejaba en dirección contraria. Después descubrí que, en vez de ir hacia la derecha como debía, iba hacia la izquierda, sin llegar a ningún sitio concreto, ya que giraba velozmente aquí y allá, mientras que la distante voz procedía en línea recta.

Deseé que al estúpido se le ocurriera tocar continuamente una cacerola de hojalata, pero no lo hizo. Lo que me preocupaba eran las pausas entre grito y grito. Bueno, continué luchando y entonces oí el grito detrás de mí. En menudo embrollo me encontraba. O bien era el grito de otra persona o yo había dado la vuelta por completo.

Dejé la pértiga. Oí de nuevo el grito; seguía a mi espalda, pero en un sitio distinto. Seguí oyéndolo, cada vez desde distinto lugar, y yo continué contestando hasta que al poco rato lo localicé nuevamente delante de mí y comprendí que la corriente había hecho dar la vuelta a la canoa, con la proa hacia río abajo, y que yo seguía la dirección correcta, si era Jim y no otro balsero el que gritaba. Era imposible distinguir una voz de otra en medio de la niebla, porque nada tiene aspecto ni sonido natural cuando hay niebla.

Prosiguieron los gritos y al minuto siguiente vi que me dirigía velozmente hacia una ribera cortada cubierta de enormes árboles de aspecto fantasmal; la corriente me arrojó hacia la izquierda y pasé rozándola entre numerosos troncos flotantes que, impulsados por la vertiginosa corriente, parecían rugir.

Un segundo o dos más tarde, se hizo todo blanco, sólido, quieto. Permanecí completamente inmóvil, escuchando los golpeteos de mi corazón, y creo que no respiré mientras dio cien latidos.

Entonces abandoné la partida. Me di cuenta de lo que ocurría. Esa ribera cortada era una isla y Jim había bajado por el lado opuesto. No era un asidero de remolque al que pudiera llegarse en diez minutos. Tenía la arboleda de una isla corriente; podía tener unas cinco o seis millas de largo y más de media milla de ancho.

Continué inmóvil, aguzando los oídos, durante unos quince minutos. Estuve flotando, claro está, a cuatro o cinco millas por hora, pero uno esto no lo piensa. No, uno lo siente mientras permanece quieto como un muerto sobre el agua; y si se vislumbra un tronco flotante que pasa por el lado, uno no piensa en lo de prisa que uno va, sino que contiene el aliento y se dice: «¡Diantre, este tronco vuela!» Si se figuran que no se siente uno solo y abandonado en medio de la niebla, en la noche, pruébenlo una vez... y ya verán.

Seguidamente, durante una media hora, grité de vez en cuando; al fin oí la respuesta, muy distante, e intenté seguirla, pero no lo conseguí, y en seguida calculé que me había metido en un nido de asideros de remolque, porque los vislumbraba vagamente a ambos lados, en ocasiones formando un angosto canal entre sí. Y otros que no podía ver y que sabía que estaban allí, porque oía el rumor de la corriente batiendo contra la maleza y la hojarasca que colgaba sobre las orillas. Bueno, no tardé mucho en perder la pista de los gritos, y sólo traté de perseguirlos un rato, de todos modos, aunque era peor que perseguir un fuego fatuo. Jamás se ha sabido de un sonido que cambiara de sitio tantas veces y con tanta velocidad.

Tuve que apartarme con las manos de la ribera en cuatro o cinco ocasiones para impedir llevarme las islas del río; y, por tanto, pensé que la balsa debía chocar contra el margen de vez en cuando, pues de lo contrario se adelantaría hasta el punto de que dejaría de oírla... pues flotaba más de prisa que yo.

Poco después me pareció encontrarme de nuevo en río abierto, pero no pude oír ni seguir el rastro de ningún otro grito. Supuse que Jim se había agarrado a un tronco flotante y que estaba perdido. Me sentía muy cansado, de modo que me tumbé en el fondo de la canoa y me dije que no me preocuparía más. No quería dormirme, claro está, pero estaba tan soñoliento, que no pude evitarlo, así que decidí echar un sueñecito.

Pero me figuro que fue algo más que un sueñecito, porque cuando desperté las estrellas refulgían, había desaparecido la niebla y me encontraba doblando un enorme recodo, con la popa por delante. De momento no supe dónde estaba; creí estar soñando. Cuando lo recordé, me pareció que todo había ocurrido la semana anterior.

Allí el río era monstruosamente grande, con árboles muy altos y espesos a ambas orillas, como una muralla sólida, según pude ver a la luz de las estrellas. Volví la mirada río abajo y vi una mancha negra sobre el agua. Me dirigí hacia allí, pero cuando la alcancé vi que se trataba de un par de troncos aserrados y atados juntos. Luego vi otra mota y la perseguí; luego otra, y esta vez acerté. Era la balsa.

Cuando me acerqué, Jim estaba sentado, con la cabeza baja y hundida entre las rodillas, dormido, con el brazo derecho colgando por encima del remo de gobierno. El otro remo estaba destrozado y la balsa cubierta de hojas, ramas y escombros. Era evidente que había pasado malos ratos.

Amarré la canoa y me tumbé en la balsa, bajo las mismas narices de Jim. Empecé a bostezar y a desperezarme golpeando a Jim con mis puños cerrados, diciendo:

—Hola, Jim, ¿me he dormido? ¿Por qué no me despertabas?

—¡Válgame Dios! ¿Eres tú, Huck? ¿No te has muerto... ni te has ahogado... y has vuelto? Es demasiado para ser verdad, querido; es demasiado. Déjame mirarte, pequeño, déjame tocarte. ¡No, no estás muerto! ¡Has vuelto sano y salvo, Huck, mi querido amigo Huck, loado sea Dios!

—¿Qué te pasa, Jim? ¿Has estado bebiendo?

—¿Bebiendo? ¿Que si he bebido? ¿Es que he tenido tiempo de beber?

—Bueno, pues entonces ¿por qué hablas como si estuvieras loco?

—¿Que hablo como si estuviera loco?

—Sí, dices tonterías sobre que he vuelto sano y salvo... ¡Ni que me hubiera ido!

—Huck... Huck Finn, mírame a los ojos, mírame a los ojos. ¿No te has ido?

—¿Quién? ¿Yo? Oye, ¿de qué diablos hablas? Yo no he ido a ninguna parte. ¿A dónde querías que fuera?

—Bueno, mira chico, aquí hay algo raro. ¡Vaya si lo hay! ¿Yo soy yo o quién es yo? ¿Estoy aquí o dónde estoy? Esto es lo que ahora quiero saber.

—Bueno, a mí me parece que estás aquí, salta a la vista, pero creo que estás bastante chiflado, Jim.

—¿Tú crees? Bueno, contéstame a esto: ¿No te llevaste el cable en la canoa para sujetarlo al asidero de remolque?

—No. ¿De qué asidero me hablas? No he visto ningún asidero de remolque.

—¿Que no lo has visto? Oye... ¿Acaso el cable no se soltó y la balsa se fue río abajo, dejándote a ti atrás, en medio de la niebla?

—¿Qué niebla?

—Pues... ¡la niebla! La niebla que ha habido toda la noche. Y ¿acaso tú no gritabas y yo grité hasta que nos metimos entre las islas, y uno de nosotros se perdió, y el otro fue como si también se hubiera perdido porque no sabía dónde estaba? ¿Y acaso yo no me estrellé contra muchas islas y lo pasé horriblemente, y pensé que iba a ahogarme? ¿No es cierto, muchacho..., no es cierto? Contéstame a esto.

—Esto ya es demasiado para mí, Jim. No he visto niebla, ni las islas, ni me ha pasado nada de nada. He estado aquí toda la noche charlando contigo hasta que te dormiste apenas hace diez minutos, y supongo que también yo me dormí. En tan poco tiempo no has podido emborracharte, de modo que seguramente lo has soñado todo.

—¡Que me cuelguen! ¿Cómo puedo haber soñado todo esto en diez minutos?

—¡Pues que me cuelguen a mí si no lo has soñado, porque nada de esto ha ocurrido!

—Pero, Huck, si para mí ha sido tan claro como...

—Esto no importa, pues nada es verdad. Lo sé porque he estado aquí todo el rato.

Jim guardó silencio durante cinco minutos, pensando. Luego dijo:
—Bueno, pues supongo que lo habré soñado, Huck, pero te juro que ha sido el sueño más real que he tenido. Y, además, ninguno me había dejado tan cansado como éste.
—Oh, es natural; a veces un sueño cansa a cualquiera. Pero el tuyo debió de ser fenomenal... Cuéntamelo, Jim.
Así que Jim empezó a relatármelo como había ocurrido, sólo que lo adornó muchísimo. Luego dijo que iba a interpretarlo, porque le había sido enviado como aviso. Dijo que el primer asidero de remolque significaba un hombre que trataría de hacernos algún bien, pero que la corriente significaba otro hombre que nos alejaría del primero. Los gritos eran avisos que nos llegarían de vez en cuando y, si no nos esforzábamos en comprenderlos, nos arrastrarían a la mala suerte en vez de alejarnos de ella. Los asideros de remolque eran dificultades que nos acarrearían las disputas con gente camorrista y toda suerte de personas viles, pero que, si nos metíamos en nuestros asuntos y no replicábamos ofendiéndolos, saldríamos con bien, abandonaríamos la niebla y entraríamos en el inmenso y claro río, que simbolizaba los Estados Unidos libres, y que entonces se acabarían las tribulaciones.
Estaba bastante nublado cuando subí a la balsa, pero empezaba a despejarse otra vez.
—Bien, hasta ahora la interpretación es bastante buena, Jim —dije yo—; pero ¿qué significan estas cosas?
Me refería a las hojas y los escombros que había en la balsa y al remo destrozado. Podían verse claramente.
Jim lo contempló, luego me miró a mí y otra vez a las hojas y lo demás. El pensamiento de que había soñado se había clavado tan fuertemente en su cerebro, que parecía incapaz de desprenderse de él y situar nuevamente los hechos en su justo lugar. Pero, cuando logró ordenar sus ideas, me miró fijamente, sin sonreír, y dijo:
—¿Qué significan? Voy a decírtelo. Cuando me quedé extenuado del esfuerzo y cansado de llamarte y me dormí, llevaba el corazón casi destrozado porque te habías perdido y no me importaba lo que se hiciera de la balsa. Y cuando desperté y te vi aquí de nuevo, sano y salvo, los ojos se me llenaron de lágrimas y me hubiera arrodillado para besarte los pies, tanta era mi gratitud al cielo. Y a ti todo lo que se te ocurrió fue reírte de mí, engañar al viejo Jim con una mentira. Esto de aquí es porquería, y porquería es la gente que mete escombros en la cabeza de sus amigos y les hace avergonzarse.
Se levantó despacio, anduvo hasta el cobertizo y entró en él sin decir nada más. Pero fue suficiente. Me hizo sentirme tan ruín, que casi le hubiera besado los pies para que retirase todo lo que había dicho.
Tardé quince minutos en decidirme a humillarme ante un negro..., pero lo hice y después no lo sentí. No le gasté más bromas mezquinas y no le habría gastado aquélla de haber sabido que iba a herirlo tan profundamente.

16

Dormimos durante casi todo el día y nos pusimos en marcha por la noche, siguiendo algo distanciados a una balsa monstruosamente larga, que pasaba tan lentamente como una procesión. En cada extremo tenía cuatro remos largos, así que calculamos que debía llevar unos treinta hombres. A bordo había cinco enormes cobertizos, separados unos de otros, una hoguera de campamento descubierta en el centro y un asta de bandera en cada punta. Era de postín. Valía mucho ser balsero de semejante balsa.

Descendimos a la deriva hasta un enorme recodo y la noche se cubrió de nubes y se hizo calurosa. El río era muy ancho, amurallado a ambas riberas por sólidas arboledas; casi era imposible ver una rendija a través o una luz. Hablamos de Cairo y nos preguntamos si lo conoceríamos cuando llegáramos allí. Yo dije que probablemente no, porque había oído decir que allí sólo había una docena de casas, y añadí que, si no tenían las luces encendidas, ¿cómo sabríamos que pasábamos junto a un pueblo? Jim replicó que, si los dos grandes ríos se unían allí, reconoceríamos el lugar. Pero yo dije que tal vez creeríamos pasar junto a una isla y que volvíamos a entrar en el mismo río de antes. Eso inquietó a Jim... y a mí también. El caso era: ¿qué íbamos a hacer? Propuse remar hasta la orilla en cuanto apareciese una luz, decir a alguien que papá iba detrás, con una chalana mercante, que tenía poca pericia en el oficio y que quería saber cuánto faltaba para llegar a Cairo. A Jim le pareció una buena idea, de modo que lo celebramos fumando una pipa mientras esperábamos.

No teníamos otra ocupación que la de mirar atentamente la ribera en busca del pueblo para no pasar de largo. Jim dijo que estaba seguro de verlo, porque sería un hombre libre en cuanto lo descubriera, y que si le pasaba por alto se encontraría de nuevo en el país de los esclavos, sin más posibilidades de obtener la libertad. De vez en cuando se levantaba de un salto diciendo:

—¡Ahí está!

No estaba allí. Eran fuegos fatuos o luciérnagas, de modo que él volvía a sentarse y continuaba vigilando como antes. Dijo que sentirse tan cerca de la libertad le producía estremecimientos febriles. Bueno, les aseguro que también yo temblaba y me sentía febril oyéndole hablar, porque empezaba a entrarme en la cabeza que Jim era casi libre... Y ¿quién tenía la culpa? Pues yo. No conseguí arrancarme eso de la conciencia. Llegó a inquietarme de tal modo, que me impedía descansar. Era incapaz de estarme quieto en un sitio. Antes no se me había ocurrido pensar en lo que estaba haciendo. Pero entonces sí lo pensé; y ese descubrimiento me perseguía, torturándome continuamente. Traté de convencerme de que yo no era culpable, por que yo no había obligado a Jim a huir de su legítima dueña, pero todo fue inútil: la conciencia se levantaba para replicar cada vez: «Pero tú sabías que huía en busca de su libertad y podías haberte dirigido remando a la orilla para decírselo a alguien». Así era... No había que darle vueltas. Ahí era donde me dolía. La conciencia me reprochaba: «¿Qué te hizo la pobre señorita Watson para que tú dejaras que su negro se fugara sin decir ni una sola palabra? ¿Qué te hizo la pobre mujer para merecer de ti un trato tan vil? ¡Vaya!, ella trató de darte una formación, de enseñarte buenos modales; intentó ser buena contigo como supo. ¡Eso es lo que ella te hizo!»

Empecé a sentirme tan mezquino y miserable, que casi deseé morirme. Me moví inquieto por la balsa, insultándome a mí mismo, mientras Jim iba de un lado a otro junto a mí. Ninguno de los dos podía estarse quieto. Cada vez que él pegaba un brinco y decía: «¡Eso es Cairo!», tenía la impresión de que me disparaban un tiro y pensaba que, si era Cairo, me moriría de desesperación.

Jim hablaba en voz alta mientras yo hablaba para mí. Decía que lo primero que haría al llegar a un Estado libre sería ahorrar dinero y no gastarse ni un centavo, y que cuando tuviera bastantes ahorros se compraría una mujer esclava de una finca cercana a la de la señorita Watson; que entonces los dos juntos trabajarían para comprar a los niños, y que si su amo no quería venderlos, ellos buscarían a un abolicionista para que se los robara.

Me dejaba helado oírle hablar así. Jamás en su vida se había atrevido a expresarse de esta forma. Fíjense en la diferencia que hubo en él tan pronto se creyó libre. Como dice el antiguo refrán: «Dale la mano a un negro y se tomará el codo.» Me dije que éste era el resultado de no pensar. Ahí esta el negro al que ayudé a huir, que entonces se desenmascaraba diciendo que iba a robar a sus hijos, los niños que pertenecían a un hombre al que yo ni siquiera conocía, un hombre que no me había hecho ningún daño.

Lamenté oírle decir eso a Jim; era algo que le desmerecía. Mi conciencia empezó a azuzarme con más violencia que nunca, hasta que le dije «Déjamelo de mi cuenta... Todavía no es tarde... Iré a la orilla remando cuando vea la primera luz y lo diré.» En seguida me sentí tranquilo, feliz y ligero como una pluma. Habían desaparecido todas mis preocupaciones. Miré atentamente en busca de una luz, casi cantando para mis adentros. Al poco tiempo apareció una y Jim me advirtió señalándola:

—¡Estamos salvados, Huck, estamos salvados! Ponte en pie y cuádrate! ¡Al fin es Cairo! Lo sé, lo sé...

—Iré a comprobarlo en la canoa, Jim. Pudiera no serlo, ¿sabes?

El se levantó y preparó la canoa; extendió su chaqueta en el fondo para que me sentara encima, me entregó el remo y, cuando yo me alejaba, gritó:

—¡Pronto estaré gritando de alegría y diré que todo se lo debo a Huck! Soy un hombre libre y nunca lo habría sido de no ser por Huck... ¡Huck lo consiguió! Jim jamás te olvidará, Huck. Eres el mejor amigo que ha tenido Jim; y ahora eres el único amigo que tiene el viejo Jim.

Me alejaba remando, animado por el deseo de denunciarlo, pero cuando él dijo eso pareció como si me abandonaran las fuerzas. Continué lentamente y no estaba seguro sobre si me alegraba de haberme puesto en camino o no. Cuando estuve a cincuenta yardas de distancia, Jim gritó:

—¡Ahí va el bueno y leal Huck! ¡El único caballero blanco que ha cumplido la promesa hecha al viejo Jim!

Bueno, es que me ponía enfermo. Pero recordé que debía hacerlo... No podía eludirlo. En aquel momento se aproximó un esquife tripulado por dos hombres armados, quienes se detuvieron, y yo hice lo mismo. Uno de ellos preguntó:

—¿Qué es aquello?

—Una balsa —contesté.

—¿Eres de esa balsa?

—Sí, señor.

—¿Hay otros hombres en ella?

—Sólo uno, señor.

—Esta noche se han escapado cinco negros, allá arriba, al otro lado del recodo. ¿Tu hombre es blanco o negro?

No contesté en seguida. Lo intenté, pero las palabras no salían. Me debatí un segundo o dos tratando de atreverme a decirlo, pero no fui lo bastante hombre... no tuve ni el valor de un conejo. Vi que me ablandaba, de modo que desistí, y afirmé:

—Es blanco.

—Iremos a verlo.

—Vayan —dije yo—; papá está allí, y ustedes pueden ayudarle a remolcar la balsa a tierra; allí donde hay luz. Está enfermo... y también lo esta mamá y Mary Ann.

—¡Oh, diantre! Tenemos mucha prisa, chico, pero debemos ir. Vamos... Empuña la pértiga y en marcha.

Yo empuñé mi pértiga y ellos sus remos. A los pocos instantes dije yo:

—Papá se lo agradecerá, se lo aseguro. Todo el mundo se aleja cuando les pido que me ayuden a remolcar la balsa a tierra, y yo solo no puedo hacerlo.

—¡Valientes sinvergüenzas! Claro que es algo raro... Oye, chico, ¿qué le pasa a tu padre?

—Es la... éste... Bueno, casi no tiene nada.

Dejaron de remar. Entonces habíamos acortado bastante la distancia que nos separaba de la balsa. Uno de ellos muy nervioso se adelantó y dijo:

—Muchacho, esto es una mentira. ¿Qué es lo que tiene tu padre? Será mejor para ti que digas la verdad.

—La diré, señor, palabra... Pero no nos dejen, se lo suplico. Tiene la...la... Caballeros, si continúan remando un poco más y me dejan echarles el cable, no tendrán que acercarse a la balsa... Háganlo, por favor...